STATISTICS
from
SCRATCH

An Introduction for Health Care Professionals

David Bowers

JOHN WILEY & SONS

Chichester · New York · Brisbane · Toronto · Singapore

Copyright © 1996 by John Wiley & Sons Ltd,
Baffins Lane, Chichester,
West Sussex PO19 1UD, England

National   01243 779777
International   (+44) 1243 779777

Reprinted December 1996, August 1997

Other Wiley Editorial Offices

John Wiley & Sons, Inc., 605 Third Avenue,
New York, NY 10158-0012, USA

Jacaranda Wiley Ltd, 33 Park Road, Milton,
Queensland 4064, Australia

John Wiley & Sons (Canada) Ltd, 22 Worcester Road,
Rexdale, Ontario M9W 1L1, Canada

John Wiley & Sons (Asia) Pte Ltd, 2 Clementi Loop #02-01,
Jin Xing Distripark, Singapore 0512

Library of Congress Cataloging-in-Publication Data

Bowers, David.
    Statistics from scratch : an introduction for health care
professionals / David Bowers.
        p.   cm.
    Includes bibliographical references and index.
    ISBN 0-471-96325-9 (pbk.)
    1. Medical statistics.   I. Title.
RA409.B67   1996
519.5'02461—dc20   95–49518
                                           CIP

British Library Cataloguing in Publication Data

A catalogue record for this book is available from the British Library

ISBN 0-471-96325-9

Typeset in 10/11½ Palatino from the author's disks by Dorwyn Ltd, Rowlands Castle, Hants
Printed and bound in Great Britain by Bookcraft (Bath) Ltd
This book is printed on acid-free paper responsibly manufactured from sustainable
forestation, for which at least two trees are planted for each one used for paper production.

STATISTICS
from
SCRATCH

10

3

CONTENTS

PREFACE

My aim in writing this book was to provide a simple introduction to statistics, which assumes no previous knowledge of the subject and avoids the use of any difficult mathematics. Indeed, the reader will need no more than simple arithmetic to understand the material included here. I have tried to write a book which uses simple everyday language, free from jargon, and which above all is user-friendly.

The focus of the material is in the area of health care and most of the examples draw on real data from this arena. However, I hope the book will also appeal to all of those in the human sciences broadly defined, e.g. in the social and behavioural sciences, and in other allied and complementary subjects.

The book covers what is generally known as descriptive statistics, i.e. how to organise, analyse and present sample data. But as well as this statistical material I have also included detailed explanations of how the main statistical computer packages (SPSS, Minitab, Excell and EpiInfo) for the analysis of data. I think to do otherwise would be like staging Hamlet without the Prince, or eating a bacon sandwich without the bacon.

I have aimed this book at both students and professionals. For those coming to statistics for the first time and for those who need a little light refreshment. Enjoy!

David Bowers
March 1996

ACKNOWLEDGEMENTS

I am very grateful to all the copyright holders who have given me permission to reproduce previously published material.

The Lancet, for Tables 1.1, 5.5 and 6.14, copyright 1994 The Lancet Ltd. The *British Medical Journal* for case material in Chapter 2 and Tables 2.1, 3.10, 3.11, 4.1, 4.3, 4.8, 5.11, 5.12, 5.13, 6.6, 6.12, 6.16 and 7.2, copyright 1993, 1994 and 1995 the British Medical Journal. HMSO for Table 3.1, Crown copyright 1995; reproduced with the permission of the Controller of Her Britannic Majesty's Stationery Office. The publishers of *Journal of Human Nutrition and Dietetics* for Tables 2.4 and 4.10, copyright 1994. The publishers of *American Journal of Occupational Therapy* for Table 3.5, copyright 1994. The publishers of *Journal of Advanced Nursing* for Tables 3.16, 3.17, 6.4, 6.5, 6.11 and 6.13, copyright 1993, 1994 and 1995. The publishers of *Community Medicine* for Table 6.15, copyright 1986.

I

GETTING STARTED

1

FUNDAMENTALS

IN THE BEGINNING . . .

Data are the numbers we get when we count or measure things—like the number of admissions to a hospital in a month, or the proportion of a GP's patients who smoke, or the opinions of nurses on a new wound dressing, or the weight of babies born to ethnic minority mothers, and so on. This book is about using *statistics* to make sense of such data. In this sense the word *statistics* describes the collection of procedures which help us to investigate the data, but it can also mean the actual numbers themselves.

Of course, people have been counting and measuring things for a very long time, but it was only the invention of writing, just over 5000 years ago, that provides us with the first hard evidence of these earliest "statistical" activities. The medium used for this first writing was clay, formed into small flat tablets and inscribed when still soft with a sharpened reed. Once it had dried the tablet provided a more or less permanent record of whatever had been inscribed on it. The method of writing which evolved from this process is now known as "cuneiform" (from the Latin *cuneus* meaning "wedge") because of the wedge-shaped indentation produced by the reed pen on the clay. Not surprisingly the idea caught on. Initially however, before the evolution of a proper written language (which took a further several centuries), these first efforts in writing were confined to the recording of numbers of various items or objects. In other words they were used to record *data*. In this sense the scribes who did the counting and recording of the data are the first statisticians whose existence we know about.

As an example, Figure 1.1 shows two very early tablets from the end of the fourth millennium BC found at Tell Brak in modern-day Syria. These tablets record "10 sheep and 10 goats"; the circular impression "O" indicates the value 10 in the early Summerian language (the value 1 would have been indicated by the smaller cup-shaped mark "▽"). We have no idea why the scribe concerned wanted to record these data (maybe as a receipt for animals bought but not yet paid for, or as a tithe paid to a local chief) but in a very real way these tablets represent a very early example of a *frequency distribution*, a very important concept in statistics which we will meet in Chapter 3.

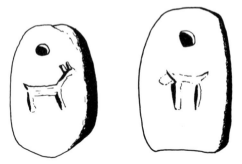

Figure 1.1:   Two early tablets from about 4000 BC, recording 10 sheep and 10 goats

It wasn't long before the scribes discovered that it was much quicker to use straight-line stylised *representations* of the items they were describing rather than draw the original realistic images. Over time these stylised representations, or symbols, became more and more abstract in appearance and less and less like the objects they portrayed. Eventually it became impossible to look at one of them and guess from its appearance alone what object it was supposed to represent. For example the word for "day" (*ud* in Summerian) started out with a comparatively realistic portrayal:

which 500 years later had become:

and 1000 years after that ended up as:

There is a message here for those of you who have previously tried to avoid statistics because of its strange symbols and off-putting language (more like double-Dutch perhaps!). Most statistics books and journal articles contain words and symbols (and ideas too perhaps) which might just as well be in Summerian. After all, it would be extremely difficult to guess that the last sign above meant "day" simply by looking at it. We would either have to be told or have a Summerian dictionary. For example the sign "Σ" looks quite like a Summerian symbol (if you hadn't seen it before) but is in fact a symbol in statistics, which means "add up" or "sum" (as you will see later).

What I'm trying to say is, don't be put off by the language of statistics, think of it as just another language which can be learnt, and without too much difficulty. In writing this book I have tried wherever possible to avoid the use of complicated and scary-looking equations, but a few have of necessity crept in. Usually you will be able to avoid them at first reading, and come back to them second or even third time round, without losing the thread. In the end I hope that you will find that statistics is a lot more accessible than perhaps you think it is right now.

FROM LITTLE TO LARGE: SAMPLES AND POPULATIONS

Suppose you were judging fruit cakes at the local village show. You would judge the cakes on the basis of a small slice taken from each. In other words you would try a *sample* of each cake and then judge the *whole* cake on the not unreasonable assumption that the small slice is typical of the whole. Exactly the same principle operates in statistics. Statisticians usually employ the data obtained from *samples* to draw conclusions about entire groups or *populations*. Here the word "population" doesn't mean what it does in school geography lessons. It is not the number of people who live in France or China or any other country (although it *could* be). A population in statistics is a *complete* collection of people, objects or items in which we have an interest. The important thing is that it contains *every* eligible person (or item) however it is defined.

MEMO

"Statistics" can mean two things:

The numbers we get when we measure and count things. We call these numbers *data*.

A collection of procedures for describing and analysing data.

Populations may be very large (even infinite in size), or quite small. It might be easy to identify every member of a given population, or as is more usual, quite difficult if not impossible. For example, it might be *all* children with asthma in the country, or all children with asthma in a certain health area, or all children with asthma registered with a certain GP, or of school age, or with parents who smoke, and so on. But *every* eligible child must be included. The notion of sampling from a population is illustrated in Figure 1.2, but I will have more to say about this in Chapter 7.

Having chosen the sample in such a way that it is *representative* (a subject we will deal with in Chapter 7), i.e. the children in the sample have similar characteristics to the children in the population, we could then with some confidence say that what we had discovered about the children in the sample is broadly true of *all* such children in the population. This process of drawing conclusions about the characteristics of populations on the basis of sample data is known in statistics as *statistical inference*.

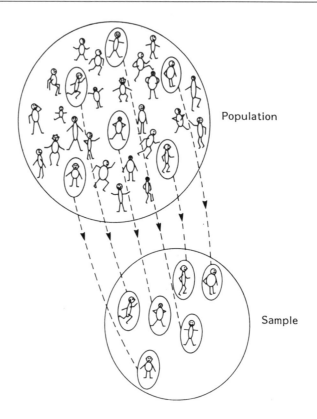

Figure 1.2: Sampling from a population

MEMO

Inference

We draw representative samples from populations and use the data from these samples to make inferences about the characteristics of the population from which the sample is drawn.

DESCRIBING OR INFERRING

In the last section I said that statisticians "usually" take samples with the aim of discovering something of the characteristics of the population from which the sample was taken. Sometimes, however, the aim might be simply to *describe* the characteristics of the sample without making any inferences about the underlying population. Thus we might have two objectives in statistical analysis: either to do no more than describe the features of the sample; or, in addition, to use the sample results to make inferences about the sampled population. Let's have a look at these two objectives in a little more detail.

Descriptive Statistics

We may only want to describe the main features of the *sample* and no more. For example, with childhood asthma, we might be interested in the average age of the children in the sample, or their weight and height, or the average length of time since their asthma was first diagnosed, the proportion whose parents smoke, etc. Such features are often referred to as the *descriptive* or *baseline* characteristics and the methods we use in obtaining them are known as *descriptive statistics*. Descriptive statistics then may include any or all of the following objectives:

- To obtain a *broad* overview of the distribution of the sample data, identifying any features and characteristics of interest which may be present. This will usually require the data to be arranged or organised into some form of table (for example putting the ages of the asthmatic children in ascending order and/ or counting how many there are of each age), and perhaps also the *representation* of the organised data by means of a chart or graph. Numeric summary values may also be determined, for example the proportion of parents who smoke.
- To determine a numeric *summary* measure of the *average* of the sample values— for example the average age of the asthmatic children, or the average duration of their illness. There will often be more than one such suitable measure from which a choice may be made. Collectively, such measures of average are known as measures of central tendency or *measures of location*.
- To determine a numeric *summary* measure of the degree to which the sample values are spread out. For example, if we are studying the incidence in children of school age, are there broadly speaking about the same proportion of children of every age from five through to 18, or are three-quarters of them aged say between 5 and 9 years, with relatively few any older? Again there will usually be more than one measure to choose from. Collectively, such measures are known as *measures of spread or dispersion*.

This book concentrates on the ideas and methods of descriptive statistics.

Inferential Statistics

We will often want to reach beyond the sample data to make inferences about the population from which the sample was taken. For example, if the average age of the children in the sample is 9.5 years, we might conclude that the average age of *all* asthmatic children in the population from which the sample was drawn is also somewhere *around* 9.5 years, give or take a few months say. In statistical inference the process we use to achieve such statements about the likely accuracy of our sample data is called *confidence interval estimation*.

Alternatively, we could use our sample data to test previously held beliefs about the average age of asthmatic children. For example, previous studies might have indicated that the average age of children of school age with asthma is 8.2 years. Does our sample average age of 9.5 years support this previously believed value or not? This approach is known as *hypothesis testing*. Figure 1.3 summarises the differences between descriptive and inferential statistics. The methods of statistical inference are dealt with in a companion book.

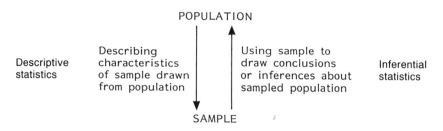

Figure 1.3: Summarising differences between descriptive and inferential statistics

---

MEMO

Statistics can be divided into:

Descriptive statistics. Used to describe the characteristics of a sample, and includes:

- Methods for organising sample data
- Methods for calculating average values
- Methods for calculating values of spread

Inferential statistics. Used to make inferences about the population from which the sample is drawn, and includes:

- Confidence interval analysis
- Hypothesis testing.

An Example from Practice

In each chapter there is at least one example, taken from real practice, which should illuminate some of the points made in the text. This first case describes the way in which a population was defined by the researchers and how a sample was taken from persons eligible to be members of that population. This enabled a set of baseline characteristics to be described.

In a study of the advantages and disadvantages of regular versus the "as-needed" use of salbutamol in the treatment of asthma[1], the eligible *population* was defined as those individuals who satisfied the following criteria: "were aged between 18 and 65; met the American Thoracic Society criteria for the diagnosis of asthma; reported requiring daily treatments with inhaled beta2-agonists; were clinically stable; and their baseline FEV1* was at least 40 per cent of predicted with a 15 per cent or greater increase 30 minutes after inhalation of 200 µg of salbutamol. Patients with significant non-respiratory illnesses and women of child-bearing potential were excluded". Using these criteria the researchers were able to recruit a *sample* of 341 asthmatic individuals from consecutive attenders at a medical clinic, for inclusion in the experiment (known as a "*trial*"). Some of their baseline characteristics are shown in Table 1.1.

---

* FEV1 is forced expiratory volume.

Table 1.1: Some baseline characteristics of
asthmatics

Proportion of male to female	123:190
Average age (years)	41.7
Pre-bronchodilator FEV1	2.08
Post-bronchodilator FEV1	2.66

Half of the individuals in the sample used their inhaler four times daily for two weeks, the other half used their inhaler as-needed. After 14 days the groups swopped over (known as a *crossover*) and the trial continued for another 14 days. Throughout, participants were asked to record the number of episodes, the number of as-needed salbutamol puffs, and their peak expiratory flow rate (a measure of their ability to breathe properly) twice a day. Subsequent inferential statistics seemed to indicate some advantages to the regular use of salbutamol over as-needed use.

USING COMPUTERS IN STATISTICS

Most of the statistical procedures which I describe in this book are much more easily performed with a computer. In fact, especially when large samples with hundreds or even thousands of observations are involved, it is difficult to conceive of many of the calculations being carried out in any other way. Personal computers are now so widespread in both office and home that I imagine that most readers of this book will have access to at least one of the major statistics or spreadsheet computer packages. For this reason I have provided fairly detailed accounts of the instructions and commands needed to perform each procedure discussed using each of four statistics packages and one spreadsheet package. There are of course many other packages which may do the job equally well, but space considerations have allowed me to include only these four:

- SPSS for Windows (Base System)[2]
- Minitab for Windows[3]
- EPI Info[4]
- Microsoft Excel

SPSS and Minitab are large general-purpose statistical packages. SPSS comes in several volumes, each increasingly more sophisticated, although only the Base System of Release 6 was used for this book. It is perhaps one of the most powerful packages of its kind.

Minitab, whilst not ultimately being quite as sophisticated and versatile as SPSS, is an extremely user-friendly package. Release 10 was used for this book.

EPI Info is a dedicated epidemiological computer package with a strong statistical element. Version 6 was used for this book. EPI is a public domain package. This means that it can be legally copied without permission and without charge. The one disadvantage of EPI Info is that, unlike the three other packages, data cannot be entered immediately into a data screen. Instead a questionnaire file and a record file have first to be created, which is inconvenient, unless of course these are already in existence for the problem in hand. EPI is a DOS-based package. I would strongly recommend that you acquire this "free" package, perhaps by borrowing the disks from an existing user (a complete and detailed manual is contained within the program and may be down-printed).

As well as these four packages, there are several others which no doubt are of equal merit or perhaps favoured by readers. These might include SPS, SAS, SMART and UNISTAT. I cannot claim to be familiar with the first three, but I have heard good things about SMART, and would single out UNISTAT as an excellent general-purpose package with a particularly good graphics program (but an extremely poor index). I would have liked to have included examples of this package in this book but space considerations meant that I had to draw the line somewhere.

Finally, I have included some examples of the Microsoft Excel spreadsheet program (Version 5). Excel is a powerful and relatively easy-to-use program in the Microsoft Office suite of programs. It contains an excellent charting facility as well as a descriptive statistics program. The range of statistical procedures in Excel can be increased significantly using an excellent "Add-in" program called "Astute"[5]. This includes a dozen or so of the most commonly used statistical procedures in the health and human sciences field.

Entering Data and Naming Variables in SPSS

Each time SPSS is loaded the opening screen displays the Output (or command) window in the upper half of the screen, and the Data Editor window containing the datasheet in the lower half. The Data Editor window is active (the active window always appears in the foreground), with the cursor in the first cell of column 1. To enter new data from the *keyboard* into this first column simply type the first data value followed by the return key* (to enter the data into any other column, click on the first cell in the desired column). After a short delay, the value you typed will appear in this first cell. Then type the subsequent values following each by the return key. These values will appear in column 1. When you have finished entering all the data, you can save it by clicking on **File** then on **Save As . . . .**

---

* "Return" key and "Enter" key are used synonymously throughout this book.

To enter data from an existing file, click on **File**, **Open**, **Data**, and then fill in the required details of the file in the Open Data File dialogue box (i.e. file name, drive, directory). When all the data are entered they can be saved by clicking on **File** then on **Save As . . .**, and then supplying the necessary details in the Save Data As dialogue box.

Each column of data (which of course corresponds to a variable) can be named by clicking on:

> **Data**
>> **Define Variable**

Then type the chosen variable name in the Variable Name box. The *type* of variable (e.g. numeric, date, currency, etc.), the number of decimal places, and any *labels* you may wish to use (for example, Male = 0, Female = 1), can be controlled by clicking on **Type . . .** and/or **Labels . . .** in the Change Settings box.

Entering Data and Naming Variables in Minitab

Each time Minitab is loaded the opening screen displays two windows, with a *sessions* or command upper window and a data or *worksheet* lower window. The data window is active with the cursor in the first cell of column 1. To enter data from the keyboard, simply type the first value and press the enter key. Repeat for the other values, pressing enter after each value (to enter data into a different column click in the first cell of the desired column). These data values will then appear in column 1 of the worksheet. To enter data from a file, click on:

> **File**
>> **Other Files**
>>> **Import ASCII Data**

Type c1 in the Store Data In Column(s) box, click **OK**, then provide location of data file in the Import Text From File dialogue box. To name a column, click in the cell immediately above the first data cell in the required column and then type the name, which—after the return key has been pressed—will appear at the head of the column.

The method I have employed to illustrate the use of each procedure is to list the commands that have to be successively clicked with a mouse (or on occasion typed), after the data have been entered, to produce the required results. For example:

> **Stat**
>> **Basic Statistics**
>>> **Descriptive Statistics**
>>>> **Select c1**
>>>>> **Select c2**
>>>>>> **OK**

means click on **Stat**, then on **Basic Statistics**, then on **Descriptive Statistics**, then click on **c1** and then on **Select**, then on **c2** and then on **Select**, and finally click on **OK**. Sometimes you will also have to click on a button (⊙) or on a check box (☒). Obviously a little trial and error may be required if you are using either a package or a procedure for the first time. I have been able to explore only a few of the many options available in the dialogue boxes of both SPSS and Minitab (particularly in the former). I would encourage you to do this when you have a few minutes to spare.

The results from any of the procedures used appears in the Output window of SPSS or the Sessions window of Minitab. Output, i.e. the results of any calculations you requested, can be saved by clicking on **File, Save SPSS Output** in SPSS, or on **File, Save Window** in Minitab, and then providing details in the dialogue box of the address of the destination file to which the output is to be saved.

Finally, I should mention that I typed this book using the Microsoft Word for Windows package, part of the Microsoft Office suite of programs. Many of the charts and diagrams included were produced using the chart program embedded within Word. Running both Word and Excel simultaneously means that charts can easily be produced in Excel, copied, and then "pasted" into a Word document (using **ALT TAB** to switch back and forth from one program to the other).

SUMMARY

This first chapter has outlined the difference between samples and populations and between inferential and descriptive statistics. Descriptive statistics confines itself to describing and summarising the main features of the sample (sometimes called the base characteristics). With inferential statistics we use these to make inferences about the characteristics of the sampled population. The ideas we have touched on are the first step in establishing a number of fundamental concepts which we will need in all that follows. In the next chapter I will describe more of these important ideas. After that we will be ready to start to use statistics with real data.

EXERCISES

1.1 What is the essential difference between descriptive and inferential statistics?

1.2 Explain what the word "population" means in statistics. A sample will never have exactly the same characteristics as the population from which it is taken. Explain.

1.3 Why do you suppose that sample data can never provide precise values for the characteristics of a population?

1.4 A nurse manager in a large general hospital wishes to investigate the prevalence of pressure sores among patients. She decides to arrange for nurse assessment of the presence and severity of pressure sores of *every* patient in the hospital at some time between 9 a.m. and 6 p.m. on a specific day.

(a) What is the population here?
(b) What factors will make it difficult to assess every patient whom medical records report as occupying a bed on that particular day?
(c) Even if every patient in the hospital on that day is assessed, why are the data collected still only a sample?

REFERENCES

1. Chapman, K. R. *et al.* (1994) Regular vs as-needed inhaled salbutamol in asthma control. *The Lancet*, **343**, 4 June.
2. SPSS® for Windows™. SPSS Inc., 444 N. Michigan Avenue, Chicago, Illinois 60611, USA.
3. Minitab® for Windows™. Minitab Inc., 3081 Enterprise Drive, State College, PA 16801–3008, USA.
4. EPI Info. Centre for Disease Control (CDC), Atlanta, Georgia 30333, USA. Supplier in the UK is the Public Health Laboratory Service, 61 Collindale Avenue, London NW9 5EQ, UK. Tel: 0181 200 6868.
5. Astute: Statistics Add-in for Microsoft Excel™. DDU Software, University of Leeds, Old Medical School, Leeds LS2 9TJ, UK.

2

TYPES OF VARIABLE

THE LONG AND THE SHORT AND THE TALL

The last chapter outlined two ways of dealing with sample data. First, we can
describe the main characteristics of a sample using *descriptive* statistics. Second,
we can use the results from the descriptive statistics procedures to investigate the
population from which the sample was taken (*inferential* statistics, dealt with in a
companion volume to this book). Chapter 3 will look at the first of these areas in
greater detail, but before that we need to consider some ideas relating to the
notion of a *variable*. The material covered in this chapter is of considerable importance for much of what is to follow in the rest of this book (not to mention statistics
as a whole!).

VARIABLES

Loosely defined, a variable is something whose value or quality can vary from
person to person, from item to item, or from moment to moment. For example:

- Height is a variable because its value can vary from one person to the next. One
  person might be 1.54 m, another 1.73 m.
- The number of empty beds in a hospital is a variable, because it can have the
  value 0, or 1, or 2, or 3, and so on.
- Blood group is a variable, it can commonly take the value A, or AB, or B, or O.
- Patients' satisfaction with their treatment is a variable because it can vary, for
  example, from very satisfied, to satisfied, to acceptable, to unsatisfied, to very
  unsatisfied.

Variables are often given abbreviated names (especially for computer use, or in
algebraic manipulation), or denoted by a single letter, very often X, which is
usually *italicised*. Thus a variable measuring levels of pain could conveniently be
abbreviated to *PAIN* or P or simply just X.

> **MEMO**
>
> A variable is a name given to anything that can take on different values or attributes. Variables come in three different types, nominal, ordinal, and metric.

We can classify any variable into one of three types, nominal, ordinal or metric, according to the sort of data used to measure its value. It is very important to be able to say what type of variable we are dealing with because the methods which we can use (in any descriptive or inferential analysis) *depend* on the variable type. Some procedures can be used with one type of variable but not with another type (as we shall see), so correct identification of variable type can be crucial if an inappropriate (and potentially misleading) approach is to be avoided. For this reason we need to consider the differences between each type of variable, which means between each type of data.

NOMINAL DATA

A variable is a nominal variable if *nominal data* are used to measure its value. Data are said to be *nominal* if:

- they consist of simple labels or categories such as male/female; or yes/no; or blood type A, type B, type O, type A/B; or employed/unemployed; or whatever;
- and the categories or labels have *no inherent ordering*, i.e. it doesn't matter whether we write male/female or female/male, or O, A/B, A, B or B, O, A/B, A, etc.—in other words, the order of the categories is completely *arbitrary*.

Thus blood group, with four categories A, A/B, B and O, is a nominal variable. The category labels we use with nominal variables are invariably alphabetic or non-numeric. Because of this we cannot use the rules of ordinary arithmetic with nominal variables. For example we can't find the "average" sex of 10 men and 20 women by adding them together and dividing by 2 (as we *could* find the average of 10 cm and 20 cm by doing the sum $(10 + 20)/2 = 30/2 = 15$ cm).

> **MEMO**
>
> *Nominal variables* (also called categorical variables) simply categorise sample values into an appropriate category. We cannot apply any of the rules of arithmetic to nominal data.

An Example from Practice

Researchers who were investigating the dating of pregnancy by ultrasound studied a sample of 7904 non-smoking and 2624 smoking mothers delivered of babies in three large maternity units[1]. Each mother was allocated to one of four ethnic categories and the number of mothers in each category (and whether she smoked or not) is shown in the accompanying table.

The variable being measured here is "ethnicity", which can take one of the four values: European, Afro-Caribbean, Indian or Pakistani, or Other.

ETHNIC GROUP	NON-SMOKERS	SMOKERS
European	7390	2541
Afro-Caribbean	169	67
Indian or Pakistani	282	3
Other	63	13

The arrangement of the ethnic categories in the table (with Europeans at the top and Others at the bottom) is *completely arbitrary*. They could have been arranged in alphabetic order, or reverse alphabetic order, or randomly, or (which is a common practice) by their increasing size, or in any other way. There is nothing about the nature of the categories to suggest that any type of ordering is either desirable or necessary. It is this arbitrariness of the category ordering that is an essential feature of nominal variables.

Even though the categories taken by nominal variables are almost always alphabetic, they will often be coded numerically for computer analysis. For example we might code smoking = 0 and non-smoking = 1; or ethnicity as European = 1, Afro-Caribbean = 2, Indian or Pakistani = 3, and Other = 4. But such coding does *not* introduce any mathematical meaning (we could equally well have coded smoking = 2 and non-smoking as 1, or used *any* two other numbers).

Nominal variables are very common in health care and so subsequent chapters discuss methods for describing and analysing nominal data. As we shall see, the fact that they only categorise data and have no inherent arithmetic content limits the sorts of analysis we can employ.

ORDINAL DATA

A variable is an ordinal variable if *ordinal data* are used to measure its value. Data are said to be *ordinal* if:

■ they consist of categories such as satisfied, undecided, unsatisfied; or stage I, stage II, stage III, stage IV (of disease); or poor, comfortable, well-off, rich; and so on;
■ and the categories have a natural or inherent *ordering* (as they do in the above examples).

So in the above example, disease stage is an ordinal variable.

An Example from Practice

Look at the data in Table 2.1. This shows[2] the responses of a sample of 37 diabetic patients to the question "Do you feel that your general practitioner makes a thorough assessment of your diabetes?"

Table 2.1: Patients' responses

PATIENT'S RESPONSE	NUMBER OF PATIENTS
Very thorough	13
Thorough	9
Adequate	12
Poor	3
Very poor	0

The variable here is "level of satisfaction with assessment" and it can take any one of the five values shown in the table, from "very thorough" (13 patients) to "very poor" (0 patients). This means we need five categories to accommodate these five different values. Each patient's response is allocated to one of these five categories and the total number of patients in each category summed.

The categories also have an inherent or natural *ordering*, the ordering shown in the table. We wouldn't think, for example, of having the categories in the order: thorough, very poor, adequate, very thorough, poor, or indeed any order other than the one shown in the table (although it would have been just as acceptable if we had started with "very poor" at the top of the table down to "very thorough" at the bottom, as long as we kept the same natural ordering in between).

So an ordinal variable is one which not only allows us to place each sample value into its proper category, but where there is also a natural ordering to those categories. Thus we can say that a patient who responds "very thorough" is *more* satisfied than a patient who only responds "thorough". However, and *this is a very important point*, we cannot say *how* much more satisfied that patient is. These are subjective judgements, which have not been measured with scientific

"It's alright dear, it's only a raw data outlet pipe."

instruments, and which in any case may change from hour to hour. Besides which one person's "thorough" might be the next's "adequate".

This means that we can't, for example, say that a patient who responds "very thorough" feels *twice* as satisfied as one who responds "thorough", or *one and a half times* as satisfied, or 25% more satisfied, etc. In other words we can't say exactly what the difference is between the categories. Thus although the categories of ordinal data can be ordered, we can't say anything about the size of the differences between the categories.

This lack of a proper arithmetic content to the categories of an ordinal variable means that, as with nominal variables, it makes little sense to add, subtract, multiply or divide ordinal data. In the last example, we can't therefore add the values in the table together and divide by 37 to find the "average" degree of satisfaction (but again, there is a way of summarising ordinal data to find a measure of its "averageness", as we shall see later). Moreover, numeric coding for computer entry doesn't turn them into proper numbers either.

MEMO

Sample data from an *ordinal* variable can not only be categorised but the categories also have a natural or inherent *order*. But we can't *quantify* (i.e. measure precisely) the difference between the categories. Consequently, as with nominal variables, we can't apply the rules of ordinary arithmetic to ordinal data.

The ordinal variables discussed above have *alphabetic* data categories. Just as common are data categories which take the form of *numbers*[*]. Three examples typical of this type of ordinal variable are, first, the many different types of rating or measuring scales[3]; second, visual analogue scales; and third, ranked scores.

Measurement or Rating Scales

There are probably thousands of different measurement or rating scales used in the health and human sciences[4]. Just a few of the more common include: the Apgar Scale (used to assess the health of newborn babies); the ISS (the Injury Severity Scale, used with trauma patients); the Pressure Sore Prediction Scale and the Waterlow Scale, for pressure sore prediction; the Barthel Activities of Daily Living (ADL) Scale (used in the assessment of an individual's ability to care for themself and lead an independent life); the Jarman and the Townsend Deprivation Scales (measures of social and economic circumstance); the Eating Attitudes Test (used in nutrition); the Glasgow Coma score; and so on.

Most rating scales require the completion of a questionnaire which will usually require either subjective judgement and/or the choice between categories whose

---

[*] But as Spock might say, "They may be numbers, Jim, but not as we know them". In other words, as you will see, these are not "proper" numbers, numbers we can do arithmetic with.

boundaries are blurred. Examples are the difference between "minor help" and "major help" (in the Barthel ADL scale) or between "often" and "very often" (in the Eating Attitudes Test). The data coming from these scales are therefore not exactly measured, and so are ordinal.

The way the questionnaires associated with these scales work will be familiar to anyone who reads popular magazines. As an example, fill in the questionnaire shown in Figure 2.1 now.

Pick the foods you eat at home and put your score at the end of every row. For example if you eat white bread put a mark of **1** in the right-hand column, if brown bread, a score of **2**, and so on. Finally add your scores up to give a total score.				
Food	**Score 1 point**	**Score 2 points**	**Score 3 points**	**Your score**
Bread	White	Brown	Wholemeal	
Breakfast cereal	Rarely or never eat cereals OR eat sugar-coated cereals	Cornflakes, Rice Crispies	Wholegrain cereals, e.g. Shredded Wheat, muesli	
Potatoes, pasta, rice	Never or very rarely eat	Eat potatoes, white rice or pasta most days	Eat potatoes in jackets, brown rice, or pasta most days	
Pulses, nuts, beans, etc.	Never or rarely eat	Less than once a week	1–3 times a week	
Vegetables (all kinds)	Less than once a week	Several times a week	Daily	
Fruit (all kinds)	Less than once a week	Several times a week	Daily	
			Total score:	

Figure 2.1: Dietary fibre questionnaire

The range of possible scores is from 6 to 18. A score of 0–7 suggests more dietary fibre is required, a score of 8–14 is good, and a score of 15–18 very good. Although we appear to have "real" numbers here, which we could perhaps use to find averages and the like, this is not the case. If you filled in the questionnaire you will realise how much subjective judgement and wavering takes place, even in a simple example like this. Somebody scoring 12 on this scale doesn't necessarily have a diet *twice* as healthy in fibre content as a person scoring six. They appear to have a *better* fibre diet perhaps, but we can't say *exactly* how much better.

MEMO

Rating scales are measures on variables, most frequently elicited by a questionnaire or by direct observation. The questions usually involve some degree of subjective judgement or inexact measurement. The numeric data they produce are, as a consequence, ordinal.

If we can't say that a score of 12 is twice a score of six, then clearly we can't do any sort of arithmetic with these data. This is why rating scales like this produce ordinal data. There is of course a lot more to devising a proper rating scale than this rather trivial example, and it is possible to find useful guides to the construction of scales if you need to devise one yourself[3].

An Example from Practice

Researchers investigating the effects of meals supplemented with either fat or carbohydrate on satiety (feeling full-up), used the Eating Attitudes Test (EAT) to ensure that the sample of 16 male subjects in the trial did not have an eating disorder[5]. The EAT scale requires the subject to respond "always", "very often", "often", "sometimes", "rarely" or "never" to 40 questions relating to food and eating. Each of the six responses carries a different score and the total score over the 40 questions provides the overall EAT score. The range of possible total scores is from 0 to 120. Low scores indicate normal eating without disorder.

The EAT scores of the 16 subjects in the trial were:

   0 4 3 0 0 0 1 0 4 0 2 1 1 1 5 7

These were judged to be low enough in every case for participation in the trial. The trial showed that carbohydrate- but not fat-supplemented meals appear to inhibit further eating.

The important point about these EAT scores is that a person with an EAT score of 4 does have a slightly less healthy attitude to food than a person with a score of 2, but not necessarily half as healthy. Nor can we say that the difference between a score of 1 and a score of 2 is the same as the difference between scores of 2 and 3 or 3 and 4, or any other pair of adjacent values.

You may find the following analogy helpful in thinking about the "fuzzy" nature of ordinal measurements. Assume that your friend Vlad gives you a seven-foot ladder whose rungs are *supposed* to be one foot apart. However, Vlad is not too good at DIY and the rungs are unequally spaced as shown in Figure 2.2. Suppose an ordinal variable X can take the values 1, 2, 3, 4, 5, 6 and 7 and these are represented by the height of each rung from the ground. So for the first rung $X = 1$ foot above the ground, for the second rung $X = 2$ feet above the ground, and so on. Now with an ordinal variable the value of 4, say, is not necessarily twice as large as the value 2, i.e. the fourth rung cannot be said to be exactly twice as high off the ground as the second rung (as indeed we see it is not). Nor is the difference between any two adjacent rungs necessarily the same as the difference between any other two adjacent rungs (as we can see).

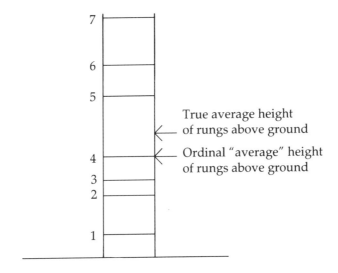

Figure 2.2: Ladder analogy for ordinal variables

If we try to find the average height of the seven rungs by adding their seven heights together and dividing by seven we get $(1 + 2 + 3 + 4 + 5 + 6 + 7)/7 = 4$ feet. But as we can see, the fourth rung of this ladder does not equal the average height of these rungs from the ground, which is in fact about half-way between the fourth and fifth rung.

Visual Analogue Scales

A second measuring device which leads to *numeric* ordinal values is the Visual Analogue Scale (VAS). This commonly takes the form of a horizontal (or vertical) line (usually 10 cm long) drawn on a piece of paper or card, such as that illustrated below:

No pain |————————————————————————————————| **Unbearable pain**

Its use by an acute pain service, for example[6], would involve asking patients to mark the point on the line which represents the level of pain they are experiencing: the left-hand end of the line representing no pain, the right-hand end, unbearable pain. A cm rule might then be used to measure the distance of the marked point from the left-hand end. Thus a patient might be given a score of 2.3 or 6.5, etc.

Although the VAS score takes the appearance of a proper mathematical measure, it is nonetheless, for all the reasons argued above, an ordinal measure. Although for any patient we can say that a pain score of 8 indicates worse pain than a score of 7, we can't say that a pain score of 8 indicates pain twice as bad as a pain score of 4. Nor can it be said that the difference in the level of pain between, say, a score of 5 and a score of 6 is the same as the difference between scores of 8 and 9. And nor can we say that a pain given a score of 5 by one individual is the same pain as that given a score of 5 by another individual.

An Example from Practice

A VAS was used in an investigation of the use of aqueous cream instead of soap to alleviate pruritus (itching) in a sample of 20 patients with liver disease[7]. The researchers stated: "Given the subjective nature of itching, a visual analogue scale (VAS) (range 1–8) was used to obtain the patient's degree of perceived itch. The VAS consisted of an eight-cm line marked 'No itching' at the left end and 'Worst possible itching' at the right. The patient was asked to mark the line at a point that corresponded to their degree of itching". The distance of the point from the left-hand end was then measured with a cm rule and this constituted that patient's score. The trial revealed the following typical scores:

	SOAP	CREAM
Typical VAS score	5.5	2.5

Since the higher the VAS score the worse the itching, these results suggest that the aqueous cream was of more use than soap in relieving pruritus symptoms.

Ranked Scores

It is sometimes appropriate (we will see why later in the book) to work with sample data that have first been *ranked* before any subsequent analysis. Ranking data is quite straightforward. For example, consider Table 2.2, the first and second columns of which contain the patient numbers and their Pressure Sore Prediction Scale (PSPS) scores, for a sample of 11 patients. PSPS is a 13-item questionnaire, completed by the appropriate health care professional, which aims to quantify the risk of a patient getting a pressure sore. The maximum theoretical score is 56 and a total score of 10 or more is indicative of pressure sore risk.

The first step is to *sort* the PSPS values into ascending order (column 3). The second step is to identify the patients corresponding to these sorted scores (column 4). The third step is to rank the scores in column 3 starting with the rank of 1 for the smallest score (4), the rank of 2 for the next smallest score (9), and so on (column 5). Alternatively, the rank of 1 can be given to the largest score, rank 2 to the next largest, and so on. It makes no difference to any subsequent analysis provided that the direction of ranking is noted. From column 5 we see that patient 9 had the lowest PSPS score (lowest risk of pressure sores), patient 1 the next lowest, and so on.

Table 2.2:   Ranking scores

PATIENT	PSPS SCORE	SORTED PSPS SCORE	PATIENT	RANKED PSPS SCORE
1	9	4	9	1
2	20	9	1	2
3	10	10	3	3
4	16	12	6	4
5	27	16	4	5
6	12	20	2	6
7	45	27	5	7
8	29	29	8	8
9	4	34	11	9
10	75	45	7	10
11	34	75	10	11

Dealing with Ties

We frequently find that two or more of the sample values are the same. For example, in Table 2.3 patients 5 and 8 now have the same PSPS score of 27. *In these circumstances the scores are given the average of the ranks they would otherwise have taken.*

Table 2.3:   Ranking scores (with ties)

PATIENT	PSPS SCORE	SORTED PSPS SCORE	PATIENT	RANKED PSPS SCORE
1	9	4	9	1
2	20	9	1	2
3	10	10	3	3
4	16	12	6	4
5	27	16	4	5
6	12	20	2	6
7	45	27	5	7.5
8	27	27	8	7.5
9	4	34	11	9
10	75	45	7	10
11	34	75	10	11

In this example patients 5 and 8 would have had their scores ranked as 7 and 8. They are thus each given the rank of 7.5 [= (7 + 8)/2]. If patient numbers 5, 8 and 11 had all had scores of 27, they would each have been given ranks of (7 + 8 + 9)/3 = 8. Regardless of how many tied values are involved, they are all given the average of the ranks they would have had if their values hadn't been tied.

MEMO

Ordinal variables can take either non-numeric or "numeric" values. Typical of the former are variables whose values result from "level of satisfaction" type questionnaires. Typical of the latter are variables measured using rating scales, visual analogue-type variables, and variables with ranked scores.

Using Computers to Sort and Rank Data

Sorting data (i.e. putting it into ascending or descending order) or ranking data, by hand, is tedious and *very* error prone if there are more than a few values. It is well worth the trouble of using a computer package such as Excel, SPSS or Minitab to do the ranking for you. We can use the above PSPS scores to illustrate the quite straightforward procedures involved.

Using Microsoft Excel to sort and rank data

Assuming that the data to be sorted are in column A, and have been selected by clicking on "A", then the following commands will sort the data into ascending order:

> **Data**
> > **Sort**
> > > **Sort by:** ⊙ **Rows**
> > > ⊙ **Ascending**
> > > > **OK**

The following commands will rank either the 11 original or the 11 sorted data if they are in column A and put the ranked values in column B:

> **Tools**
> > **Analysis Tools**
> > > **Ranks and Percentiles**
> > > > **(Input range:) A1:A11**
> > > > > **(Output range:) B1:B11**
> > > > > > **OK**

Using SPSS to sort and rank data

The PSPS scores are entered into column c1 of the SPSS datasheet and the variable named PSPS following the procedure described in Chapter 1. To sort the data first follow the commands:

> **Data**
> > **Sort Cases**
> > > **Select PSPS**
> > > > **OK**

The original data column is replaced by the sorted values. To rank either the original or these sorted values, the required commands are:

> **Data**
> > **Rank Cases**
> > > **Select PSPS** (i.e. either the original or the sorted scores)
> > > > **OK**

The ranked scores then replace the sorted (or original) scores. The sorted and ranked scores are shown in Figure 2.3.

```
SPSS for MS Windows Release 6.0

    4
    9
   10
   12
   16
   20
   27
   27
   34
   45
   75

    1.000
    2.000
    3.000
    4.000
    5.000
    6.000
    7.500
    7.500
    9.000
   10.000
   11.000
```

Figure 2.3:   The PSPS scores from Table 2.2, sorted and ranked using SPSS

Using Minitab to sort and rank data

The PSPS scores are entered into column c1 of the Minitab worksheet and named PSPS following the procedures described in Chapter 1. To sort the data follow the commands:

Manip
   Sort
      Select c1
         type **c2** in the Store sorted column(s) box
            type **c1** in the Sort by column box
               OK

To rank the data the commands needed are:

Manip
   Rank
      Select c2
         type **c3** in the Store ranks in box
            OK

The resulting output is shown in Figure 2.4.

PSPS	SortPSPS	RankPSPS
9	4	1.0
20	9	2.0
10	10	3.0
16	12	4.0
27	16	5.0
12	20	6.0
45	27	7.5
27	27	7.5
4	34	9.0
75	45	10.0
34	75	11.0

Figure 2.4: The sorted and ranked PSPS scores using Minitab

Health care statistics frequently involves work with nominal and ordinal variables. The fact that we can't use ordinary arithmetic with this sort of data limits the ways in which we can present, summarise or analyse it. Fortunately, however, we will just as frequently have to deal with variables which produce data which we *can* add, subtract, multiply and divide. It is to such variables that we now turn.

METRIC DATA

A variable is a metric variable if *metric data* are used to measure its value. Data are said to be *metric* if:

■ they consist of precisely defined values, such as 36°C, 37°C, 38°C, 39°C, etc., or 3000 g, 3001 g, 3002 g, etc., which can be placed in exact order;
■ and the difference between *any* two *adjacent* values is the same as the difference between any other two adjacent values.

In the above example, temperature and babies' weight are both metric variables. Metric data are also known as interval/ratio data (the word "numeric" is also used by some authors).

Metric data are "real" numbers. We can add, subtract, multiply and divide the sample values, and so find averages, and other numeric measures. So, for example, we could calculate the average weight of two babies weighing 3.2 kg and 4.4 kg as (3.2 + 4.4)/2 = 3.8 kg. Unlike nominal and ordinal data, metric variables have *units of measurement*, e.g. weight (in mg, g or kg), time (in seconds, minutes, days, weeks, etc.), temperature (in °C), counts of things (e.g. numbers of patients, admissions, operations), and so on. The data recorded by the Summerian scribes those thousands of years ago (see Figure 1.1) were metric, consisting of a count of things, the numbers of sheep and goats.

In terms of the ladder analogy, metric data values are like the evenly spaced rungs on a properly made ladder, as shown in Figure 2.5, where each rung is now a correct measure of height from the ground.

The ordinal ladder on the left might represent the Apgar scores of seven babies. A baby with an Apgar score of 6 is not twice as healthy (twice as high off the ground

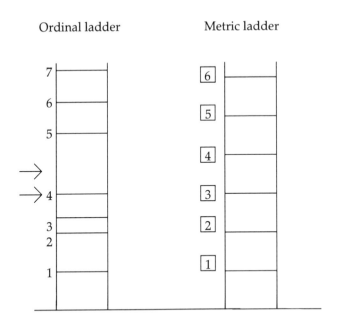

Ordinal ladder   Metric ladder

Figure 2.5:   Ordinal and metric variables compared as two ladders

in the analogy) as a baby with an Apgar score of 3. Nor is the difference between any two Apgar scores (rungs), say 2 and 3, necessarily the same as the difference between any other two adjacent scores, say 3 and 4. The metric ladder on the right might represent the number of babies born in one day in a small maternity unit. Six babies is exactly twice as many (twice as high off the ground) as three babies, and the difference between any two adjacent numbers of babies, say 2 and 3, is the same as the difference between any other two adjacent numbers of babies, say 3 and 4.

Stretching this ladder analogy to breaking point, the categories of a nominal variable would be rather like the rungs of a ladder lying flat on the ground. It wouldn't matter which rung you stood on, you would still be the same height from the ground. In other words the order of the rungs is then arbitrary.

As a simple example of metric data, consider the following waiting time (in minutes) before their first seeing a doctor by each of five patients in an outpatient unit:

20   25   27.5   35   40

In this situation we know that the five-minute difference between the first and second patients is *exactly* the same as the five-minute difference between the fourth and fifth patients. Moreover, we also know that a time of 40 minutes is exactly twice as long as a time of 20 minutes, and we can add the five waiting times together and divide by five to get an average waiting time. In other words, we can employ all the methods of mathematics with metric sample data.

MEMO

A metric variable (also called a numeric or an interval/ratio variable) is measured with data that consist of real numbers and to which we can apply all the rules of mathematics. Metric variables always have *units of measurement* associated with them.

As a consequence of this proper mathematical property, the choice of methods for presenting, summarising and analysing metric variable data is wider and ultimately more powerful, as we shall see.

DISCRETE VERSUS CONTINUOUS VARIABLES

As well as the distinction between nominal, ordinal and metric, variables can also be classified as being *discrete* or *continuous*. *Discrete* variables are those whose data have a limited number of values or categories, and there are *no other* possible values in between these values. Nominal and ordinal variables are inherently discrete. For example, it is possible for a patient to have a Waterlow Pressure Sore Risk Scale score of, say, 6 or 7, but not a score between 6 and 7. Moreover, the number of different possible scores is limited, to 21 (from 0 to 20).

Metric variables, however, can be either discrete or continuous. Discrete variables usually come from *counting* things, such as numbers of patients, numbers of nurses, numbers of smears taken, etc.*, and again the number of possible values is limited. On the other hand, *continuous* metric variables can in theory take an *unlimited* number of values within a given range. For example, a patient's temperature could be 31°C or 31.1°C or 31.11°C or 31.111°C, *ad infinitum*. We can always squeeze another number between any two adjacent numbers. The actual *realisable* number of possible values is limited in practice by the accuracy of our thermometer or clock, or tape measure or weighing scales, etc. Common examples of continuous variables are those involving height, weight, time, volume, temperature, pressure, density, and so on.

MEMO

- *Nominal* and *ordinal* variables are inherently discrete.
- *Metric* variables can be either discrete or continuous.
- A *discrete* variable can take only a limited number of possible values.
- A *continuous* variable can take an unlimited number of values.

---

* Note that a discrete variable which can take only two different values is known as a *dichotomous* variable. Sex is a dichotomous variable for example, since it can only be male or female. Similarly, variables which allow only "yes" or "no" answers are dichotomous.

Both the ladders shown in Figure 2.5 are of course discrete (they each have a limited number of values (rungs). If you are on one of the ladders you are on one of the rungs, not in between them. A continuous ladder would in theory have an infinite number of rungs.

MEMO

■ Discrete data usually come from counting things.

■ Continuous data usually come from measuring things.

An Example from Practice

In the study already referred to, on the appetite suppressing properties of fat and carbo-hydrate meal supplements[6], the basic characteristics of the first five members of the sample are shown in Table 2.4.

Table 2.4: Dietary supplements study

PATIENT NUMBER	AGE (years)	WEIGHT (kg)	HEIGHT (m)	BMI (kg/m^2)	EATING ATTITUDES TEST	EATING INVENTORY TEST	MEAN ENERGY INTAKE (kJ)
1	22	66.8	180.0	20.6	0	1	11 109
2	22	85.3	188.0	24.1	4	1	13 293
3	22	77.0	185.4	22.4	3	6	10 083
4	23	65.0	174.5	20.2	0	3	8 297
5	19	73.0	170.0	25.2	0	1	13 016

Each subject in the study was measured for the value of the seven variables shown. "Patient number" is a discrete nominal variable. "Age" is a continuous metric variable, despite its discrete appearance; the researchers were happy to have it measured only to the nearest whole year (on the other hand, age last or next birthday would be discrete). "Weight", "Height" and "BMI" are also continuous metric variables, each measured to just one decimal place*. The "Eating Attitudes Test" and the "Eating Inventory Test" are both discrete ordinal variables (they are both rating scales). Finally "Mean energy intake" is a continuous metric variable, measured to the nearest whole kJ. Notice that all the metric variables have units attached to them, unlike the nominal and ordinal variables. Recall that having units is a characteristic of metric variables.

It is worth reiterating that identifying a variable as being nominal, ordinal or metric is of great importance if the appropriate methods of analysis are to be used. Whether a variable is discrete or continuous can be equally as important. The strongest statistical procedures (those which produce the most reliable results)

---

* BMI stands for Body Mass Index, a measure of body size. It is equal to weight (in kg) divided by height (in metres) squared. A BMI over 30 indicates obesity.

can only be applied to the analysis of metric variables. Weaker (less reliable) methods have to be used with ordinal variables; and the least powerful (least reliable) techniques with nominal variables.

"There was no doubt that Albert knew the difference between discrete and continuous."

QUALITATIVE AND QUANTITATIVE DATA

A final distinction in types of data which is of interest is that between *qualitative* data and *quantitative* data. Qualitative data we collect when we measure or count nominal and ordinal variables. It may be alphabetic or "numeric", but does not have any true mathematical properties. Quantitative data are collected from metric variables. Such data are always numeric and have proper mathematical properties.

MEMO

- Qualitative data are used with nominal and ordinal variables.
- Quantitative data are used with metric variables.

SUMMARY

In this chapter we have seen that there are three main variable types, nominal, ordinal and metric. Nominal variables can only categorise. Ordinal variables not only categorise but the categories can also be ordered. Metric variables are measured with real numeric data. We can apply the rules of ordinary arithmetic (and, for example, therefore calculate average values) only to metric data. Although ordinal and nominal data can sometimes take "numeric" form, these values are not "real" numbers, and it is not appropriate therefore to add, subtract, multiply or divide them (this means, for example, that we can't find the average of a set of ordinal scores). All nominal and ordinal variables are inherently discrete.

Metric variables can be either discrete or continuous. Discrete variables can take only a limited number of values. Continuous variables can take, in theory, an unlimited number of values. Finally we noted that qualitative data are collected from nominal and ordinal variables, and quantitative data from metric variables.

The ideas, the language, and the concepts we've looked at in Chapters 1 and 2 are fundamental to the whole of statistics. But so far, it's as if we've built a hospital and furnished it with beds and equipment, but have not yet dealt with any patients. We are now going to start doing just that. In the remainder of this book you will see how what you have learnt so far can be used to describe, summarise and analyse health care data.

EXERCISES

2.1 The ladder analogy was used here to explain the difference between nominal, ordinal and metric variables. Can you think of any other (better) analogies?

2.2 If you were trying to explain the difference between discrete and continuous variables to a group of people new to statistics, you might use the number of eggs in an egg carton, and a kitchen weighing scale of the clock type (i.e. a pointer moving round a circular scale). Can you think of any other suitable everyday analogies that might be used to illustrate the difference?

2.3 Give two examples of each of the following variable types: (a) metric continuous; (b) metric discrete; (c) non-numeric ordinal; (d) "numeric" ordinal; (e) non-numeric nominal; (f) "numeric" nominal.

2.4 The following variables (and their possible values) were measured by researchers comparing effectiveness of treatments for painful shoulder syndrome (PSS). Identify each variable as nominal, ordinal or metric (and whether discrete or continuous):

Sex (M, F)
Age (years)
Duration of PSS (months)
Number of sessions (1, 2, 3, . . .)
Range of flexion (degrees)
Pain (cm scale)
Diagnosis (tendenitis, or bursitis, or peri-arthritis)

2.5 In research on periodontal disease and mortality, data on the following variables (and their possible values) was collected. Identify each variable as nominal, ordinal or metric (and whether discrete or continuous):

Sex (M, F)
Age (years: 25–34, 35–44, etc.)
Caucasian (Y, N)
Education (high school, high school graduate)
Currently married (Y, N)
Smoking (never, former, current, unknown)

Diabetes (Y, N)
High blood pressure (Y, N)

2.6 In research into the relationship between parental neglect and obesity in young adulthood, data were collected on the following variables (and their possible values). Identify each variable as nominal, ordinal or metric and whether discrete or continuous:

Parents (two biological, single biological, biological plus step-parent, other guardian, not known)
No. of siblings (0, 1, 2, ≥ 3)
Perceived parental support (harmonious, overprotective, modest, none, not known)
Child's hygiene (well-groomed, averagely groomed, dirty and neglected, not known)

2.7 In research into the effects of passive smoking on women in the workplace, Chinese statisticians measured the following variables (and their possible values). Identify each variable as nominal, ordinal or metric (and whether discrete or continuous):

Age (years)
History of hypertension (Y, N)
Type A personality (Y, N)
Total cholesterol (mg/dl)
High-density lipoprotein cholesterol (mg/dl)
Passive smoking from husband (Y, N)
Passive smoking at work (number of smokers: 0, 1, 2, 3, . . .)

2.8 The accompanying data give the number of GPs per 100 000 of the population in 12 countries. Arrange the countries in rank order with the country with the most GPs given rank 1.

France	147
Belgium	127
Ireland	63
UK	54
Switzerland	58
Norway	66
Italy	102
Denmark	67
Hungary	78
Portugal	61
Netherlands	43
Spain	50

2.9 In a study referred to previously[6] the 16 patients in the trial achieved the scores in the Eating Attitudes Test (EAT) and the Eating Inventory Test (EIT) shown in the accompanying table. Convert both variable scores into rank order.

Subject	1	2	3	4	5	6	7	8	9	10	11	12	13	14	15	16
EAT	0	4	3	0	0	0	1	0	4	0	2	1	1	1	5	7
EIT	1	1	6	3	1	1	1	0	6	3	3	2	6	1	7	8

REFERENCES

1. Wilcox, M. *et al.* (1994) Birthweights from pregnancies dated by ultrasonography in a multicultural British population. *BMJ*, **307**, 588–91.
2. Hurwitz, B. *et al.* (1993) Promoting the clinical care of non-insulin dependent (type II) diabetic patients in an inner city area: one model of community care. *BMJ*, **306**, 624–30.
3. Davis, B. A. and Bush, H. A. (1995) Developing effective measurement tools: a case study of the Consumer Emergency Care Satisfaction Scale. *Journal of Nursing Care Quality*, **9**(2), 26–35.
4. McDowell, I. and Newell, C. (1987) *Measuring Health: A Guide to Rating Scales and Questionnaires.* Oxford: OUP.
5. Cotton, J. R. *et al.* (1994) Dietary fat and appetite. *Journal of Human Nutrition and Dietetics*, **7**, 11–24.
6. See, for example, Ferrell, Betty *et al.* (1995) Pain and quality assessment/improvement. *Journal of Nursing Care Quarterly*, **9**(3), 69–85.
7. Ronanyne, C. *et al.* (1993) The use of aqueous cream to relieve pruritus in patients with liver disease. *British Journal of Nursing*, **2**, 10.

II

DESCRIPTIVE STATISTICS

"I think that's how John gets
some of his results."

ORGANISING QUALITATIVE DATA

EVERY SET OF DATA TELLS A STORY

We saw in Chapter 1 that *describing* the sample data set means:

- arranging or organising it so that we can get a broad overview of any features which might be of interest
- finding some measure of the average of the sample values
- finding some measure of how spread out the sample values are.

Essentially we want the data to tell us their "story". We will look at measures of average in Chapter 5 and measures of spread in Chapter 6. This present chapter describes ways of organising and presenting qualitative data (i.e. data from nominal or ordinal variables). Chapter 4 will do the same for quantitative data.

I am going to leave discussion of the various methods we might use to gather data until we have examined the various methods available for describing it in this and the next three chapters. If this feels a bit like putting the cart before the horse, I should explain that I feel that it is time to get down to some examples of handling data. In any case, many readers will already be familiar with simple methods of data gathering (from questionnaires for example) or will perhaps already have a collection of data and want to know what they can do with it. Readers anxious to understand more about the practicalities of *getting* data can read Chapter 7 first without any loss of continuity and then come back to this chapter.

ORDER OUT OF CHAOS

When sample data are first gathered together from the various pieces of paper or files on which they are initially set down (e.g. from completed questionnaires, patients' notes, measurement scale results, case notes, drug records, etc.) they are inevitably disordered. Transferring the data to a single document or computer file provides a first opportunity to check for any missing, unusual or unexpected

Table 3.1: Reasons given by a sample of patients for their failure to attend an outpatient clinic appointment

PATIENT NUMBER	REASON FOR NON-ATTENDANCE	SORTED REASONS FOR NON-ATTENDANCE (ALPHABETIC ORDER)
1	Childcare problems	Childcare problems
2	Transport problems	Childcare problems
3	Time off work problems	Childcare problems
4	Time off work problems	Childcare problems
5	Forgot	Childcare problems
6	Wrong day/time	Family problems
7	Unwell on day	Forgot
8	No reason given	Forgot
9	Childcare problems	Forgot
10	Forgot	Forgot
11	Treatment thought unnecessary	Forgot
12	Time off work problems	Forgot
13	Transport problems	Forgot
14	Unwell on day	Forgot
15	Overslept	Forgot
16	Forgot	Forgot
17	Time off work problems	Forgot
18	Forgot	Forgot
19	Time off work problems	Funeral
20	Childcare problems	No reason given
21	Unwell on day	No reason given
22	Funeral	No reason given
23	Treatment thought unnecessary	No reason given
24	No reason given	On holiday
25	Time off work problems	Other medical appointment
26	Family problems	Overslept
27	Forgot	Transport problems
28	Forgot	Transport problems
29	Time off work problems	Transport problems
30	No reason given	Transport problems
31	Wrong day/time	Transport problems
32	Forgot	Time off work problems
33	Transport problems	Time off work problems
34	Unwell on day	Time off work problems
35	Child care problems	Time off work problems
36	Transport problems	Time off work problems
37	Transport problems	Time off work problems
38	Forgot	Time off work problems
39	Time off work problems	Time off work problems
40	On holiday	Time off work problems
41	Unwell on day	Treatment thought unnecessary
42	Forgot	Treatment thought unnecessary
43	Forgot	Treatment thought unnecessary
44	Other medical appointment	Unwell on day
45	Time off work problems	Unwell on day
46	Forgot	Unwell on day
47	Treatment thought unnecessary	Unwell on day
48	No reason given	Wrong day/time
49	Forgot	Wrong day/time
50	Childcare problems	Wrong day/time

values. Some of these might be simple errors of transcription, others might need further checking (this process is known as *data screening* or *data cleaning*). Eventually we will have a "clean" set of data, but in this unorganised or "raw" state ("raw" because, like raw sewage, it hasn't been treated in any way) it will be difficult to gain any insights into the data.

One obvious improvement is to organise the data into alphabetic or numeric order. This has the effect of gathering all identical categories or values together. This simple step on its own can be very illuminating. Look at the raw data in Table 3.1 for example, which shows the reasons given by a sample of 50 patients for their failure to attend an outpatient clinic appointment[1]. These data are qualitative since reason for non-attendance is clearly a nominal variable.

Because this sample is relatively small it is perhaps not as difficult to get an overall view of the reasons for non-attendance as it would have been with a larger sample. Even so, it is considerably easier to see what's going on if we sort the data into alphabetic order (gathering like-categories together), as shown in the last column of Table 3.1 (which we can do with either SPSS or Minitab as explained in Chapter 2). Scanning down the sorted data we get the immediate impression that three reasons (the categories "forgot", "difficulty getting time off work", and "not well on the day") seemed to account for a major proportion of the failures to attend.

But even with Table 3.1 we still have to do a manual count of each category if we want to get a more accurate picture. The next step is to organise the data into a *frequency distribution*, which puts it into a form from which any patterns in the data are much more readily observed, and which also provides a basis for any charts we might subsequently wish to draw, remembering the old (but true) cliché that a picture conveys a thousand words. We may also choose to calculate some numeric measure, such as a ratio or proportion. We will examine these various possibilities using, wherever possible in this book, real data.

FREQUENCY DISTRIBUTIONS

A *frequency distribution* is simply a list of the categories or values that a variable can take, together with the number of values in each category. The information is usually presented in the form of a table. The first column, containing the different categories or classes, is usually labelled with the variable name (and if metric with the units of measurement). The second column records the number of values (known as the frequency) in each category or class. This frequency column is usually labelled with a description of whatever is being counted—for example, number of patients, number of admissions, number of episodes, and so on. The concept of a frequency distribution is of great importance in statistical analysis, and the term will be used often throughout this book.

MEMO

A frequency distribution is a list of the categories or values that a variable can take together with a count of the number of items in each category.

Incidentally, the Summerian scribe's frequency distribution (see again Figure 1.1) is shown in Table 3.2. The frequency distribution has only two categories, but of course we don't know what other types of animals or items might have been counted on unrecovered tablets.

Table 3.2: Frequency distribution for the types of animals recorded by the Summerian scribe

CATEGORY (type of animal)	FREQUENCY (number of animals)
Sheep	10
Goats	10
Total	20

An Example from Practice

We start with the raw "did not arrive" (DNA) data in Table 3.1. Examination of the sorted data reveals that the variable has 13 different categories. However, there are four patients who gave no reason and five other categories which contain only one patient. If we lump these together we get seven specific categories and an "other/no reason given" category. This means we need eight categories or rows for the first column of the frequency distribution shown in Table 3.3. The second, the frequency column, is the number of patients giving each particular reason, and is determined by simply counting the number of times in the sample each of the reasons for non-attendance occurs.

Table 3.3: Frequency distribution for data in Table 3.1

REASON FOR NON-ATTENDANCE	NO. OF PATIENTS
Forgot	12
Time off work problems	9
Childcare problems	5
Transport problems	5
Unwell on day	4
Wrong day/time	3
Treatment thought unnecessary	3
Other	9
Total	50

The frequency distribution in Table 3.3 is thus one possible way of *organising* the data. It achieves the first objective of descriptive statistics in that we get a much clearer picture of the broad pattern of the DNA reasons than is possible from the raw data. We can immediately see, for example, that "forgot" and "problems getting time off work" are much the largest categories, accounting for 12 and 9 of all 50 reasons respectively. Two other problems are moderately important, child-care and transport problems. This information is not easily obtained from an examination of the raw data (and in practice many samples will be much larger).

Relative Frequency

It is often more illuminating to present a frequency distribution with the frequency of each category or class expressed as a *percentage of the total frequency*. Such frequency distributions are called *relative* or *percentage* frequency distributions. They are particularly useful if we want to compare two or more distributions when the samples are of different size. To transform the frequency of any given category to relative frequency, we divide by total frequency and multiply by 100:

$$\text{Relative frequency} = \frac{\text{frequency}}{\text{total frequency}} \times 100\%$$

Transforming the frequency distribution in Table 3.3 in this way produces the relative frequency distribution of Table 3.4.

Table 3.4: Relative or percentage frequency distribution for data in Table 3.3

REASON FOR NON-ATTENDANCE	NUMBER OF PATIENTS	PERCENTAGE OF PATIENTS
Forgot	12	24.0
Time off work problems	9	18.0
Childcare problems	5	10.0
Transport problems	5	10.0
Unwell on day	4	8.0
Wrong day/time	3	6.0
Treatment thought unnecessary	3	6.0
Other	9	18.0
Total	50	100.0

We can see now that 24% of patients said they forgot, 18% said they had problems getting time off work, and so on.

Using a Computer to Get Frequency Distributions

Using SPSS to get frequency distributions

The reason for non-attendance data are entered into column 1 of the SPSS data sheet using the value labels 1 = forgot, 2 = work, etc. from the **Data, Define Variable, Labels** commands. To produce the frequency distribution the following commands are used:

> **Statistics**
> > **Summarize**
> > > **Frequencies**
> > > > **Select c1**
> > > > > **OK**

SPSS produces the output shown in Figure 3.1.

```
SPSS for MS WINDOWS Release 6.1

DNA    Reason

                                                          Valid      Cum
Value Label               Value   Frequency  Percent    Percent    Percent

Forgot                      1        12        24.0       24.0       24.0
Time off work               2         9        18.0       18.0       42.0
problems
Childcare problems          3         5        10.0       10.0       52.0
Transport problems          4         5        10.0       10.0       62.0
Feeling unwell on           5         4         8.0        8.0       70.0
day
Wrong day/time              6         3         6.0        6.0       76.0
Treatment unnecess          7         3         6.0        6.0       82.0
Other/No reason             8         9        18.0       18.0      100.0
given

                                    ----                  -----     -----
Total                                 50                 100.0      100.0

Valid cases 50    Missing cases 0
```

Figure 3.1:   Frequency distribution produced by SPSS for patient non-attendance data

The "Value" column lists the numeric variable labels used. The "Frequency"
column counts the values. The "Percent" column gives the relative or percentage
frequencies. The "Valid Percent" column takes into account missing values, if any
(ignore the last column for the moment).

Using Minitab to get frequency distributions

Minitab can be used to produce frequency distributions but, unlike SPSS, cannot
as easily deal with alphabetic data*. Thus we could input the outpatient data
coded as 1 = forgot, 2 = time off work problems, etc. into column c1 and produce a
frequency distribution using the commands:

    Stat
        Tables
            Tally
                Select c1
                  ⊠ **Percents**
                  OK

The frequency distribution is shown in the first two columns of Figure 3.2, and the
relative frequency distribution in column 4 (ignore the third column for now). The
problem with this output is that we have to remember which reason for non-
attendance each numeric code refers to.

Using EPI to get frequency distributions

EPI can also be used to produce a frequency distribution. If the outpatient DNA
data are entered into the field REASON of a data file called "dna.rec" say, then
the commands:

---

* See the **Convert** command in Minitab Help.

```
MTB > Tally ''Reason'';
SUBC >   Counts;
SUBC >   CumCounts;
SUBC >   Percents;
SUBC >   CumPercents.

Summary Statistics for Discrete Variables

Reason       Count       CumCnt      Percent
   1           12           12        24.00
   2            9           21        18.00
   3            5           26        10.00
   4            5           31        10.00
   5            4           35         8.00
   6            3           38         6.00
   7            3           41         6.00
   8            9           50        18.00
   N =         50

MTB >
```

Figure 3.2: Frequency distribution produced by Minitab for patient non-attendance data

Program
   Analysis
      read dna.rec
         freq REASON

will produce the required frequency distribution.

Grouped Frequency Distributions

The frequency distributions above had one row for each category. In this case, when each category in the frequency distribution has its own row (i.e. when there are as many rows as there are categories) we call the frequency distribution an *ungrouped* distribution. There is no rule about the correct number of rows for a frequency distribution but somewhere between 5 and 15 is about right. If there are many different categories, and thus *too many* rows in the frequency distribution, it becomes difficult to see any patterns that might exist in the data. This defeats the object of the exercise.

When the qualitative data are non-numeric, the number of possible categories or values is usually small enough to allocate one row to each category (as we did with the DNA data). With numeric qualitative data this will also often be the case.

An Example from Practice

The Glasgow Coma Scale has only 12 values, the Apgar scale only 10. Table 3.5 illustrates the use of an ungrouped frequency distribution with a sample of Glasgow Coma scores from a sample of 28 adults with traumatic brain injury who had been in the community for at least eight months prior to the study[2]. (Note: 13–15 = mild injury; 9–12 = moderate injury; ≤8 = severe injury. The symbol ≤ means *less than or equal to*.)

Table 3.5: A frequency distribution for numeric qualitative data

GLASGOW COMA SCALE SCORE	FREQUENCY (No. of adults)
3	3
4	2
5	5
6	7
7	5
8	6
Total	28

In fact this frequency distribution requires only six rows since there are only six different values in the sample.

However, with some numeric ordinal variables the number of possible values may be quite large; e.g. the Injury Severity Score (ISS) has a possible 75 values, and the Barthel ADL Scale (measuring the degree of independent functioning), 100 values. In the latter case, this means that potentially any corresponding frequency distribution could have 100 rows! This would cover several pages and would hardly help in uncovering any features or patterns of interest. Moreover, with a sample of 50 at least half of these classes would be empty and therefore redundant.

In these situations it is helpful to *group* the values together into *classes* to make the number of rows in the frequency distribution a manageable size. We call such frequency distributions *grouped* distributions. For example we might group a sample of ISS scores as: 0–9, 10–19, 20–29, and so on, up to 70–79. Notice that it is important that the classes don't overlap, i.e. classes of 0–10, 10–20, etc. In this situation it would be uncertain into which of two possible classes a sample value of 10 should be placed.

MEMO

With some numeric qualitative data the number of possible values is too large to allocate one row to each category or value in the frequency distribution. In these circumstances the data can be *grouped* into a smaller, more manageable number of classes.

An Example from Practice

We can illustrate this with the raw data in Table 3.6 relating to the change in the mobility of patients receiving physiotherapy for arthritis. The table records scores obtained by patients on the Oswestry Mobility Scale, before and after physiotherapy (the higher the score, the better the mobility of the individual concerned). This scale is ordinal and thus produces qualitative data.

Table 3.6: Raw Oswestry Mobility Scale scores for physiotherapy patients: flexion plus extension scores before and after treatment

PATIENT	BEFORE	AFTER
1	35	70
2	60	65
3	15	75
4	55	55
5	115	115
6	90	110
7	80	85
8	100	120
9	10	15
10	10	15
11	10	55
12	10	40
13	80	20
14	25	30
15	10	55
16	95	100
17	25	30
18	110	132
19	63	70
20	112	90
21	10	40
22	72	75
23	73	83
24	60	65
25	100	105
26	100	100
27	43	40
28	90	95
29	110	105
30	70	85
....	....	....
....	....	....

Table 3.7: Ungrouped frequency distribution for the Oswestry scores in Table 3.6

SCORE	BEFORE	AFTER
10	6	0
11	0	0
12	0	0
13	0	0
14	0	0
15	1	1
16	0	0
17	0	0
18	0	0
19	0	0
20	0	1
21	0	0
22	0	0
23	0	0
24	0	0
25	2	0
26	0	0
27	0	0
28	0	0
29	0	0
30	0	2
31	0	0
32	0	0
33	0	0
34	0	0
35	1	0
36	0	0
37	0	0
38	0	0
39	0	0
40	0	3
....	....	....
....	....	....

We could organise the data into an ungrouped frequency distribution as in Table 3.7. The rows from 0 to 9 are omitted, since no patients have scores in this range, and only the first 40 of the remaining values are shown; but clearly, because there are 122 different possible values, the resulting frequency distribution would be very long with, as can be seen, lots of zero frequencies. This does not help at all.

However, if the values are *grouped* into classes as in Table 3.8 the number of rows in the before and after frequency distributions can be reduced to a more informative 14. Each of these classes is 10 wide (the *width* of a class is the distance from the bottom of one class to the bottom of the *next higher* class, e.g. from 10 to 20, from 20 to 30, and so on).

Table 3.8: Grouped frequency distribution for
the Oswestry scores in Table 3.6

SCORE	BEFORE	AFTER
0–9	0	0
10–19	7	1
20–29	2	1
30–39	1	2
40–49	1	3
50–59	1	4
60–69	3	2
70–79	3	4
80–89	2	3
90–99	3	2
100–109	3	4
110–119	4	2
120–129	0	1
130–139	0	1

Table 3.8 is much more revealing than either the raw data of Table 3.6 or the ungrouped frequency distribution of Table 3.7. We can see immediately, for example, that the commonest score "before" physiotherapy is between 10 and 19 (seven patients), whereas the commonest score "after" physiotherapy is higher and shared between four patients with a score between 50 and 59, four between 70 and 79, and four between 100 and 109. Moreover, the "after" scores cover a slightly wider range of values (i.e. are more widely spread) than the "before" scores, although the distributions of the scores are otherwise fairly similar.

This example shows that by organising the sample data into a grouped frequency distribution we can radically improve our awareness of any broad patterns that might be present. However, we would still be hard pressed to say whether or not the "after" scores are, on average, significantly higher than the "before" scores, or to quantify the difference, if any, in the spreads of the scores. These are problems we will consider again in Chapters 5 and 6. We will also return to a more detailed examination of the construction of grouped frequency distributions in the context of quantitative data in the next chapter.

MEMO

In a grouped frequency distribution try to ensure that:

- Classes are of equal width
- Classes don't overlap
- There are between 5 and 15 classes.

Using a Computer to Get Grouped Frequency Distributions

Using EPI to get grouped frequency distributions

In EPI, if the ungrouped Oswestry scores are held in the field "oswescore" (say) of the file "oswest.rec", then they can be grouped into a variable, called say "oswesgrp", with classes each 10 units wide, with the following:

> **Programs**
> > **Analysis**
> > > (type) **read oswest.rec**
> > > > (type) **define oswesgrp** _____
> > > > > (type) **recode oswescore to oswesgrp by 10**
> > > > > > **list oswesgrp**

Using SPSS to get grouped frequency distributions

With SPSS, if the raw data is in column 1, and we want to put the grouped values into a new column, we can use the rather lengthy series of commands:

> **Transform**
> > **Recode**
> > > **Recode into Different Variable**
> > > > **Select c1**
> > > > > **Name Output Variable** (e.g. Grouped)
> > > > > > **Change**
> > > > > > > **Old and New Values**
> > > > > > > > ⊙ **Range** (enter range, e.g. 0–9)
> > > > > > > > > ☒ **Output Variables are Strings**
> > > > > > > > > > **New Value** (enter new value, e.g. 0–9)
> > > > > > > > > > > **Add**

Continue this process until all required groups have been recoded, then click on **Continue** and **OK**. The grouped values will then appear in the column headed Groups. The process is quite lengthy but, once practised, reasonably simple. The other possibility with SPSS is to get the program to do the grouping for you. This procedure allows no control over the width of the classes but is more convenient. If the data are in column 1 say, then the necessary commands are:

> **Statistics**
> > **Summarize**
> > > **Explore**
> > > > **Select c1** (for Dependent List)
> > > > > ○ **Both**
> > > > > > ⊙ **Statistics**
> > > > > > > ○ **Plots**
> > > > > > > > **Statistics . . .**
> > > > > > > > > ☐ **Descriptive**
> > > > > > > > > > ☒ **Grouped Frequency Table**
> > > > > > > > > > > **Continue**
> > > > > > > > > > > > **OK**

Using Minitab to get grouped frequency distributions

With Minitab use the commands:

Manip
    Code Data Values

and then respond to the queries in the dialogue box to produce the required grouped values.

Cumulative Frequency Distributions

Consider the data in Table 3.9 (ignore the last two columns for a moment), which shows the frequency distribution of Disability Rating Scale (DRS) scores for the same sample of 28 adults who had suffered traumatic brain injury referred to above[2]. The DRS is used to measure level of disability and has a range of possible values from 0 (none) to 12 (severe). The guideline DRS scores are: 0 = no disability; 2–3 = partial; 4–6 = moderate; 7–11 = moderately severe.

Table 3.9:   Disability Rating Scale scores for 28 adults following traumatic brain injury

DRS SCORE	NUMBER OF PATIENTS	PERCENTAGE OF PATIENTS	CUMULATIVE FREQUENCY	PERCENTAGE CUMULATIVE FREQUENCY
0	1	3.6	1	3.6
1	9	32.1	10	35.7
2	2	7.1	12	42.8
3	5	17.9	17	60.7
4	5	17.9	22	78.6
5	3	10.7	25	89.3
6	0	0	25	89.3
7	0	0	25	89.3
8	2	7.1	27	96.4
9	1	3.6	28	100.0

If we ask the question "How many of the adults had a DRS score of 5?", we could answer immediately "three". But if we are asked "How many had a DRS score of *less* than 4?", the question could not be answered without some adding up (i.e. 5 + 2 + 9 + 1 = 17). The word used in statistics for this adding up is "cumulating" and if we do this same sum for every category we get what is known as a *cumulative* frequency distribution.

We get cumulative frequency by adding up the values in the frequency column, category by category, starting at the top of the column*. The last two columns of Table 3.9 show the cumulative and relative cumulative frequencies found by cumulating the frequency and relative frequency columns respectively. We can now immediately answer any questions of the type "How many, or what percentage of, adults had DRS scores of less than (or below, or under, etc.) any of the

---

* Cumulative frequencies are not appropriate for nominal frequency distributions because of the arbitrary category ordering.

possible values?" For example, 25 (or 89.3%) of adults had DRS scores less than 6 (i.e. they were less than "moderately severely disabled").

Cumulative frequency distributions *can* be used with non-numeric ordinal data, but are more often employed with numeric ordinal data or (as we will see in the next chapter) metric data. Of course it makes no sense to use cumulative frequency with nominal data because of the arbitrary category ordering.

MEMO

Cumulative frequencies (found by adding up the column of frequency values from the top) are useful for answering questions of the "How many less or fewer than?" type.

Using Computers to Get Cumulative Frequency

Figure 3.1 shows that SPSS produces relative cumulative frequency values automatically when frequency distributions are requested. EPI Info will also produce cumulative counts automatically with the **freq** command. The Minitab **tally** command will do the same if the **cumulative counts** and/or **cumulative percents** boxes in the Tally dialogue box are checked (see Figure 3.2). Microsoft Excel produces a percentage cumulative frequency distribution automatically with the Descriptive Statistics program (see Figure 5.9).

CHARTING QUALITATIVE DATA

Frequency distributions are usually very helpful in enabling us to see what's going on in the data, but it often helps to display the same information in *chart* form. This will often provide further insights into features and patterns in the data which may perhaps have previously been overlooked. Besides which, a chart can often deliver a much more immediate message, without the possible distraction of the detailed information in a frequency distribution. We can chart both ungrouped and grouped frequency distributions. There are several different chart types to choose from.

Bar Charts

If the data are qualitative and in a small enough number of categories (which will often mean non-numeric) then a bar chart is appropriate. Again there is no rule for the maximum number of categories that can be plotted with a bar chart, but too many will defeat the purpose of providing information effectively. Bar charts may be "simple", displaying only one variable, or "clustered", displaying several variables (but charting too many variables on one graph is likely to obscure the basic rationale of the bar chart, i.e. to give a clear picture of the essential features of the data).

Conventionally, a bar chart has frequency (or relative frequency) on the vertical axis, and the categories on the horizontal axis. One bar is drawn for each category, with a height equal to its frequency; all bars have the *same* width; and all are separated by *equal* spaces. The width and spacing are a matter of choice (constrained of course by the size of the page), and the number of categories that a bar chart can display is limited by the same consideration. The space between the bars is a reminder that bar charts are used with *discrete* data.

MEMO

- A bar chart is a graphical representation of one or more freuqency distributions.
- Bar charts are best used to display *qualitative* data with a smallish number of categories.
- The height of each bar equals the frequency of the corresponding category or value.
- The bars are all the same width and are equally spaced.
- With nominal data, the relative positioning of the categories on the horizontal axis is arbitrary, but they are often ordered in terms of decreasing frequency values.

Two Examples from Practice

Figure 3.3 shows the simple bar chart for the outpatient non-attendance, data based on the grouped frequency distribution in Table 3.3.

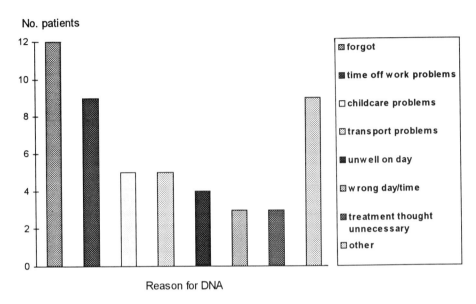

No. patients

Reason for DNA

Figure 3.3: Simple bar chart for the outpatient non-attendance data, based on the grouped frequency distribution in Table 3.3

The message delivered by the frequency distribution is confirmed by this bar chart which emphasises the importance of just two main reasons for non-attendance.

As an example of a clustered bar chart, Figure 3.4 shows the scores obtained on the Social Activities and Distress Scale by 59 adult clients at the start and end of therapy received for child sexual abuse[3], derived from Table 3.10.

Table 3.10:   Scores obtained on the SADS

SCORE	START	END
0–4	2	10
5–8	2	5
9–12	8	6
13–16	6	10
17–20	15	9
21–24	14	10
25–28	12	9

The chart shows quite clearly that on the whole the SADS scores are higher at the beginning of therapy than at the end, indicating that the therapy was successful.

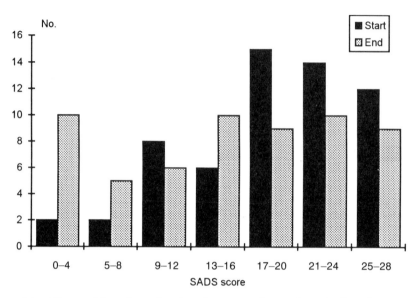

Figure 3.4: Clustered bar chart showing the scores obtained on the Social Activities and Distress Scale (SADS) by 59 adult clients at the start and end of therapy received for child sexual abuse (normal score < 9)

There is a third form of bar chart called the *stacked* bar chart, in which the bars representing the frequencies of the different categories are stacked, not alongside each other as with the clustered bar chart, but on top of each other. However, stacked bar charts are often confusing and difficult to interpret, and in general should be avoided. For this reason I am not going to discuss them.

Pie Charts

Pie charts provide an alternative to the simple bar chart, and will be familiar to most readers. The angle at the centre of each "slice" of the pie (out of a total angle of 360°) equals the frequency of its respective category compared to total frequency. Pie diagrams and bar charts are both effective ways of illustrating the main features of sample data, the choice between them is often subjective. However, a pie chart may be more appropriate if the frequency of each category is to be compared to the *total* frequency, rather than to the frequencies of the *other* categories, for which a bar chart might be preferred. The principle limitation of a pie chart is that it can represent only *one* variable (although of course several pie charts can be used to represent several variables).

An Example from Practice

To illustrate the use of pie charts we start with the raw data in Table 3.11 which contains data on the change in their professional relationships experienced by a sample of 35 GPs in a large health centre following their becoming fundholders[4]. The variable "Change in

Table 3.11: GP assessment of change in professional relationships as a result of fundholding (Key: agd = a great deal; qal = quite a lot; tse = to some extent; al = a little; naa = not at all)

GP	WITH CONSULTANTS	WITH OTHER DOCTORS	WITH PATIENTS
1	agd	qal	tse
2	qal	al	naa
3	tse	tse	naa
4	naa	naa	naa
5	naa	al	naa
6	tse	naa	naa
7	al	naa	naa
8	tse	al	naa
9	agd	qal	qal
10	qal	tse	al
11	al	naa	naa
12	agd	agd	tse
13	tse	qal	al
14	qal	tse	al
15	agd	qal	tse
16	al	naa	naa
17	naa	naa	naa
18	naa	naa	naa
19	naa	naa	naa
20	al	naa	naa
21	tse	tse	naa
22	tse	naa	naa
23	qal	al	naa
24	qal	tse	naa
25	qal	naa	naa
26	tse	naa	naa
27	agd	naa	naa
28	qal	naa	naa
29	naa	naa	naa
30	al	naa	naa
31	qal	al	naa
32	naa	naa	naa
33	tse	al	naa
34	tse	al	al
35	al	al	al

Table 3.12: Frequency distribution from Table 3.9

DEGREE OF CHANGE	WITH CONSULTANTS	WITH OTHER DOCTORS	WITH PATIENTS
A great deal	5	1	0
Quite a lot	8	4	1
To some extent	9	5	3
A little	6	8	5
Not at all	7	17	26

relationship" is non-numeric ordinal and the data are thus qualitative. Table 3.12 shows the corresponding ungrouped frequency distributions based on these raw data.

For example, in terms of the change GPs experienced in their relationships with consultants, the angle for "A great deal" is equal to: $(5/35) \times 360° = 51.4°$.

The pie chart illustrating the change in relationships with consultants is shown in Figure 3.5.

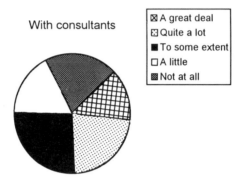

Figure 3.5: Pie chart illustrating the change in relationship of GPs with consultants following fundholder status

---

MEMO

- Pie charts are an alternative to the simpler bar chart.
- The size (i.e. the diameter) of a pie chart is arbitrary.
- The angle of each slice of the pie is proportional to the frequency of the corresponding class in the frequency distribution.
- Each pie chart can only display one variable.

Charting Cumulative Frequency

Most qualitative ordinal data that has "numeric" form comes from measurement and rating scales of various sorts (and is inherently discrete). The Oswestry Mobility Scale scores in Table 3.6 and the Disability Rating Scale (DRS) scores in Table 3.9 are good examples of this. It is sometimes useful to chart cumulative frequency distributions for such data. These charts take the form of a *step chart*, in which cumulative frequency is plotted on the vertical axis against values of the variable on the horizontal axis. Consider the DRS data in Table 3.9 again, the last column of which is % cumulative frequency. The corresponding step chart is shown in Figure 3.6.

In a step chart, the height of each step from the horizontal axis represents the cumulative or percentage cumulative frequency (whichever is being plotted), and the height of each individual step represents the frequency or percentage frequency of that value. For example, we can see that just over 60% of these subjects had a DRS score of 3 or less.

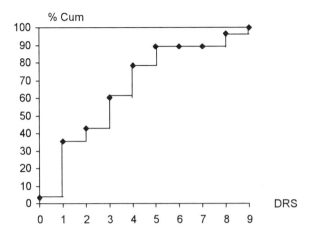

Figure 3.6: Step chart of percentage cumulative frequency for Disability Rating Scale scores for data in Table 3.9

Chapter 2 mentioned that ordinal data are inherently discrete in the sense that all we can say about any ordinal value is that it is lower than, the same as, or higher than any other ordinal value. There is no underlying continuity in the values. However, some ordinal scores, for example visual analogue scores, masquerade as being continuous, because they often have decimal values. As an example, the data in Table 3.13 show the grouped VAS percentage scores recorded by a sample of chronic migraine sufferers in response to the following question one hour into their next attack: "Compared with the worst pain you have ever suffered during a migraine attack, how would you rate the pain you are now experiencing?".

Table 3.13: VAS scores recorded by migraine sufferers one hour after start of an attack

VAS SCORE	PERCENTAGE OF PATIENTS	CUMULATIVE PERCENTAGE OF PATIENTS
0.0–0.9	6	6
1.0–1.9	9	15
2.0–2.9	16	31
3.0–3.9	24	55
4.0–4.9	20	75
5.0–5.9	11	86
6.0–6.9	7	93
7.0–7.9	4	97
8.0–8.9	2	99
9.0–9.9	1	100

With data such as these it would be more practical to use not a step chart but a *cumulative frequency curve* or *ogive*. An ogive is charted by plotting the cumulative frequency of each class on the vertical axis, against the lower value of the next higher class; e.g. with the migraine pain data, 6 against 1, 15 against 2, etc. The first point is plotted at the origin, 0 against 0. The points are then joined with a smooth curve. The result is shown in Figure 3.7.

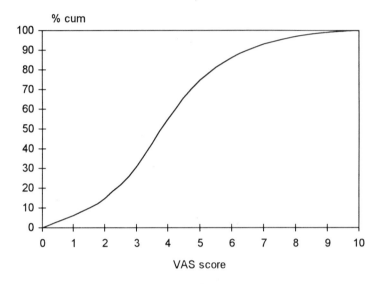

Figure 3.7: Cumulative frequency curve or ogive for cumulative percentage of VAS pain scores (from data in Table 3.13)

We can use the ogive to answer questions such as: "What percentage of subjects had a VAS score of less than 5?". Reading from a value of 5 on the horizontal axis up to the curve and then across to the vertical axis gives a value of 75. Alternatively we can see that 25% of the subjects had a VAS score of less than about 2.5. Ogives are useful for displaying cumulative continuous frequencies graphically.

Line Charts

Line charts are widely used in descriptive statistics for the display of data, particularly if the data have a chronological basis, i.e. if the data consist of regular measurements taken every minute, day, week, year, etc. Time, for example, is plotted on the horizontal axis and the frequency of one or more categories on the vertical. Such time-based data are often referred to as *longitudinal* or time-series data. The data we have examined up to this point are known as *cross-sectional* data, and are the value of a variable at some *fixed* point or interval in time.

An Example from Practice

In a study of recent trends in hospital admission rates for asthma in the East Anglian Health Region[5], researchers presented their data partly in the form of a line chart (Figure 3.8). The frequency data on the vertical axis are the number of admissions of children aged from 5 to 14 for every 10 000 children in the population in that age group. Male and female categories are plotted separately for comparison purposes. Male and female are of course nominal data categories. The chart provides for a much more immediate impression of the relative changes in the male and female admission rates than does the numeric frequency distribution. The rising trends in asthma admission rates for both males and females, with male rate rising faster and for longer, are clearly seen. Of course if we want actual numbers of admissions then these could be added to the chart.

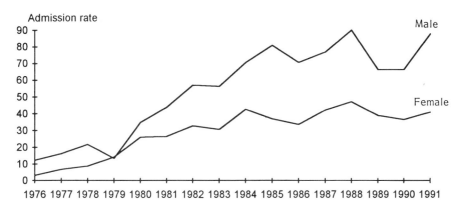

Figure 3.8:   Annual number of hospital admissions for asthma of children aged from 5 to 14, for every 10 000 children in East Anglia between 1976 and 1991

Using a Computer to Produce Charts

Using Microsoft Word and Excel to produce charts

All of the charts in this chapter were produced using the charting facility included in the *Word for Windows* word-processing package. The *Excel* spreadsheet package will also produce charts. In either case the basic procedure is quite similar and straightforward. In Word, the table containing the data to be charted is selected (or columns of the table, if the whole table is not required) using the mouse to click and drag. In Excel, the columns of data in the spreadsheet which are to be charted are similarly selected. Then the **Chart** icon is clicked on. In Word, the Chart Wizard will immediately open the Step 1 dialogue box. In Excel, the mouse should be clicked anywhere on the screen, the left mouse button held down and the mouse dragged to form a rectangle in which the Excel chart will appear, then the mouse button released. The Chart Wizard works through a five-step program as follows.

Step 1: Confirms range of cells to be charted: Click on **Next** (if correct).

Step 2: Select a chart type: For example, for a vertical bar chart, click on **Column** and then on **Next**. For a horizontal bar chart, click on **Bar** and then on **Next**.

Step 3: Select a chart format: Choose the exact form of the chart type chosen in Step 2. Then click on **Next**.

Step 4: Sample chart with options: A sample chart is displayed and a number of options are presented. For example:

Data Series in: ○ Rows
                 ⊙ Columns

Use First   Column for Category (X) AxisLabels

Use First   Row(s) for Legend Text

Make appropriate choices and click on **Next**.

Step 5: More options:

Add a legend: ○ Y
               ⊙ N

Chart Title:

Axis Titles: Category [X]
              Value   [Y]

After all choices have been made, click on **Finish**.

If you want to change the appearance of a chart after you have created it, you can do so by double-clicking on it. This will reload the Chart program. There are many details of the chart that are impossible to change, and you will need to spend some time experimenting with the wide range of possibilities for yourself. However, here are some examples:

- To change the chart type, after you have double-clicked on the chart and selected the whole chart by clicking on it, then click on **Format, Chart type**, and then on the new chart type, e.g. **Pie**.
- To change the size and type of font of the axis text (e.g. the values of X or Y), click on any of the axis text, then on **Format, Font**, and then choose the required font type and size before clicking on **OK**.
- To change the colour of the bars (or segments of the pie), click on any one of the bars, then on the colour icon, then on the required colour, then on **OK**.
- To change the background colour, click anywhere on the chart background, then on the colour icon, then choose the required colour and click on **OK**.
- To make the bars overlap or change the spacing between them, click on any one of the bars, then on **Format, Chart Type, Options**, then change **Overlap** or **Gap Width**, for example, then click on **OK**.

Many other word-processing packages (e.g. WordPerfect and Claris Works) offer similar charting facilities. Otherwise, Harvard Graphics is a more sophisticated dedicated graphics package which offers many different types of graphical display, and I have already mentioned the superior charting capability of UNISTAT.

Using SPSS to produce charts

In SPSS, charts of various types can be obtained by clicking on **Graphs**. This produces a drop-down menu of different chart types, including **Bar, Pie**, etc. Clicking on the required chart type gives the dialogue box for the selected chart. For example, to produce a clustered bar chart, click on **Clustered** and complete the dialogue box. Then, on clicking **OK** the chart appears in the chart carousel. To modify the chart, click on **Edit**. For further instructions, see the SPSS *User's Guide*.

Using Minitab to produce charts

To draw charts using Minitab, click on the **Graph** menu. For bar charts choose **Chart** . . . (bar charts are the default). For pie charts choose **Pie Chart** . . . . The corresponding dialogue boxes can then be completed and the required chart produced. You should consult the Minitab *Reference Manual* for further information.

Which Presentation Method is Best?

Most qualitative data can be charted as either a bar, pie or line chart. Pie charts are most often used for presentational purposes, but the choice as to which is most appropriate will depend on the particular circumstances.

We have examined a number of ways of organising qualitative data. These methods are not mutually exclusive; i.e. we can, if it seems appropriate, organise data into both a frequency distribution and a chart. A frequency distribution offers a more precise view of the data (since there are numbers to look at), a chart offers a more immediate impressionistic view. If the results of the investigation are to be presented at a meeting or included in a document, then a chart is often a very effective way of communicating the main thrust of the data. For those interested in reading more about the good and bad ways of presenting data in tables and charts, Myra Chapman's book *Plain Figures* offers many valuable guidelines[6].

OTHER NUMERIC MEASURES: PROPORTIONS AND RATIOS

We will often want to know what proportion of the total frequency is accounted for by one or more of the categories, or perhaps the ratio of one category to another. The calculation of either of these values is straightforward.

Proportions

The proportion of total frequency taken by any one category is:

Category proportion = frequency of category divided by total frequency

For example, in the outpatient non-attendance data (Table 3.3), we might be interested in calculating the proportion of all patients who gave "Forgot" as their reason for non-attendance:

Proportion of patients who forgot = 12/50 = 0.24

If we multiplied these values by 100 we would of course obtain relative frequency (in %). So proportions are percentages divided by 100.

Ratios

If we want to compare the frequency of one category, not with total frequency, but with the frequency of *another* category we can use the *ratio*. The ratio of the frequency of one category to another is written as:

Frequency of one category : frequency of the other category

and is usually expressed in the form:

  1 : something   or   something : 1

The former is achieved by dividing the right-hand side value by the left-hand side value, and the latter vice versa.

For example, the ratio of patients who forgot to those who had problems getting time off work is given by:

  Ratio of "forgot" to "time off work problems" = 12 : 9 = 12/9 : 1 = 1.33 : 1

and

  Ratio of "time off work problems" to "forgot" = 9 : 12 = 1 : 1.33

That is, for every one patient who had trouble getting time off work, one and a third patients said they forgot.

MEMO

- Proportions are used to compare the frequency of a category with the *total* frequency.
- Ratios are used to compare the frequency of one category with that of another category.

SUMMARY

Sample data usually consist of a largish number of unorganised values, often on several variables. Data in this form are sometimes known as raw data. Organising raw data means rearranging them so that their principal features (those of interest to the observer) are more clearly revealed. There are two principal ways of organising qualitative data, either as a frequency distribution or as a chart.

A frequency distribution is an extremely important concept in statistics. It consists of a list of the categories taken by a variable, and the number of values (i.e. the frequency) in each category. A relative frequency distribution expresses the frequencies in percentage terms.

A bar chart displays the frequency distribution of one or more qualitative variables graphically. Columns or bars of equal width and spacing are drawn so that the height of each is equal to the frequency of the corresponding category. Bar charts provide a view of the overall shape of the distribution.

Pie charts are most often used to display qualitative data when the number of categories is not too large. We need one pie chart for each variable. Line charts are used to present data, particularly that which has a chronological basis.

Proportions are used to compare the frequency of a specific category to total frequency. Ratios are used to compare the frequency of one category to the frequency of a different category.

In this chapter we have looked at ways of organising and charting qualitative data, i.e. data associated with nominal and ordinal variables. In the next chapter we turn to the organisation (and charting) of quantitative data.

EXERCISES

3.1 The raw data in Table 3.14 represent part of the results of a pressure sore audit with elderly patients at a large district general hospital. The Waterlow Pressure Sore Risk Assessment Scale is a 10-item questionnaire, which provides a measure of the risk of developing pressure sores. The maximum possible score is about 50; a score of 10 indicates some risk of a pressure sore; a score of 15 or more indicates high risk; and a score of 20 or more a very high risk.

(a) Identify the measurement status (nominal, ordinal, or metric) of each variable.

(b) Construct relative frequency distributions (grouped if necessary) for: (i) gender; (ii) age group; (iii) support surface; and (iv) Waterlow score. Comment on any broad patterns or features noticeable in each case.

(c) Display the four frequency distributions determined in (b) as: (i) simple bar charts; and (ii) pie charts. Comment on the usefulness of the charts.

(d) Calculate the proportion of (i) males, (ii) females, with pressure sores.

(e) Determine the ratio of males to females with pressure sores.

3.2 (a) Draw a clustered bar chart to display the "before" and "after" grouped frequency distributions for the Oswestry mobility scores in Table 3.8.

(b) Draw as smooth a line as possible to join the tops of all the "before" bars; repeat for the "after" bars. Comment on what is revealed about the distribution of mobility scores in these two samples.

(c) Now go back to the raw data in Table 3.6 and reconstruct the grouped frequency distributions using classes of 5–14, 15–24, 25–34 (and so on) instead of the classes used in Table 3.8. Repeat parts (a) and (b) and comment on any differences in your results.

3.3 (a) Construct (i) a relative frequency distribution, and (ii) a relative cumulative frequency distribution for the Glasgow Coma Scale data in Table 3.5. Comment on what is revealed by the former, and use the latter to estimate the percentage of individuals with a score of less than 6.

(b) Chart the data with a step chart and confirm the result obtained in (a).

3.4 Table 3.15 contains information on the computer-calculated risk of further bleeding for a sample of 4510 patients in 21 countries admitted to medical centres with upper gastrointestinal bleeding. The percentages of patients who did rebleed and who died is also given.

(a) Calculate the relative frequency distribution for the percentage of patients rebleeding.

(b) Draw a clustered bar chart showing the three relative frequency distributions.

(c) There are two ways to draw most clustered bar charts. Draw the alternative to the format you used in (b). Which do you prefer, if any, and why?

Table 3.14: Waterlow scores from a pressure sore prevalence audit (Key: column 1 = patient number; column 2 = gender; column 3 = age group; column 4 = support surface (mattress type): 01 = ordinary mattress, 02 = Vaperm, 03 = Spenco, 09 = other; column 5 = number of pressure sores; column 6 = Waterlow score)

PATIENT NO.	SEX	AGE GROUP	SUPPORT SURFACE	NO. OF SORES	WATERLOW SCORE
1	M	87–97	02	1	22
2	M	65–75	01	0	13
3	M	65–75	01	0	9
4	F	65–75	02	0	18
5	F	76–86	09	0	19
6	F	76–86	02	1	20
7	F	76–86	02	0	21
8	M	65–75	02	0	14
9	M	65–75	02	0	16
10	M	65–75	02	0	13
11	F	76–86	09	0	15
12	F	65–75	09	0	16
13	F	76–86	09	0	9
14	F	76–86	09	0	15
15	F	87–97	03	4	27
16	F	65–75	02	0	25
17	F	87–97	03	1	23
18	F	65–75	09	4	31
19	F	65–75	01	1	16
20	F	65–75	01	0	9
21	F	87–97	03	2	24
22	F	76–86	01	0	11
23	F	65–75	03	0	11
24	F	65–75	03	1	14
25	F	65–75	01	0	6
26	F	76–86	03	2	14
27	F	65–75	01	3	23
28	F	65–75	01	0	16
29	F	65–75	01	2	17
30	F	65–75	01	0	12
31	F	65–75	01	0	10
32	F	65–75	01	0	14
33	F	76–86	03	0	18
34	F	65–75	01	0	15
35	M	65–75	03	1	17
36	F	65–75	01	0	11
37	F	65–75	03	0	12
38	F	65–75	01	0	6
39	F	65–75	01	0	13
40	M	65–75	01	0	6

Table 3.15:   Risk of rebleeding after upper gastrointestinal bleeding admission

COMPUTER RISK CATEGORY	NO. OF PATIENTS	PERCENTAGE WITH REBLEEDING	PERCENTAGE WHO DIED
Very high	562	63.8	30.0
High	182	47.8	15.9
Medium	650	21.8	4.9
Low	256	12.5	1.6
Very low	973	4.0	0.2

3.5 The data in Table 3.16 show the responses of samples of traditional student nurses and Project 2000 student nurses to the question: "The nurse's role is dominated by physical care?"[7]. Represent the data with (i) a clustered bar chart, (ii) pie charts. Compare the two ways of presenting the data.

Table 3.16:   Responses of traditional and Project 2000 student nurses to the question "The nurse's role is dominated by physical care?"

STUDENTS	STRONGLY AGREE	AGREE	NOT SURE	DIS-AGREE	STRONGLY DISAGREE
Traditional	5	24	1	13	0
Project 2000	2	13	6	20	2

3.6 The data in Table 3.17 are reconstructed from a study[8] into the turnover rates of staff in the health service by length of service (turnover rate is the proportion or percentage of staff who quit their job in a given time period). Chart the data using (i) a clustered bar chart, (ii) pie charts, (iii) a line chart. Comment upon what is revealed. Justify any preferred chart format for these data.

Table 3.17:   Turnover rates of staff in the NHS by length of service (numbers are percentage of total sample quitting)

LENGTH OF SERVICE (years)	FULL-TIME NURSING	PART-TIME NURSING	FULL-TIME NON-NURSING	PART-TIME NON-NURSING
Under 1 year	18.5	22.0	21.8	28.5
1–2 years	23.5	25.2	22.1	20.0
2–3 years	26.0	22.0	20.0	12.8
3–4 years	19.6	18.7	23.0	12.3
4–5 years	20.0	15.8	15.8	11.0
5+ years	14.0	10.0	9.9	8.1

REFERENCES

1. National Audit Commission (1995) *National Health Service: Outpatient Services in England and Wales, 1994–95.* London: HMSO.
2. Hallett *et al.* (1994) Role change after traumatic brain injury in adults. *American Journal of Occupational Therapy*, **48**(3), 241–6.
3. Smith, D. *et al.* (1995) Adults with a history of child sexual abuse: evaluation of a pilot therapy service. *BMJ*, **310**, 1175–8.
4. Newton, J. *et al.* (1993) Fundholding in Northern region: the first year. *BMJ*, **306**, 375–8.
5. Hyndman, S. J. *et al.* (1994) Rates of admission to hospital with asthma. *British Medical Journal*, **308**, 1596–600.
6. Chapman, M. (1986) *Plain Figures.* London: HMSO.
7. Mitchinson, S. (1995) A review of the health promotion and health beliefs of traditional and Project 2000 student nurses. *Journal of Advanced Nursing*, **21**, 356–63.
8. Gray, A. M. and Phillips, V. L. (1994) Turnover, age and length of service: a comparison of nurses and other staff. *Journal of Advanced Nursing*, **19**, 819–27.

ORGANISING QUANTITATIVE DATA

❏ WAYS OF ORGANISING DATA ❏ FREQUENCY DISTRIBUTIONS AGAIN ❏ GROUPED FREQUENCY DISTRIBUTIONS ❏ CHARTING FREQUENCY DISTRIBUTIONS: HISTOGRAMS AND POLYGONS; OGIVES ❏ FREQUENCY CURVES ❏ THE NORMAL CURVE ❏

NEVER MIND THE QUALITY, FEEL THE QUANTITY

We get *quantitative* data when we measure *metric* variables (recall that metric data always have units attached). We organise and chart quantitative data using much the same methods as for qualitative data. The most useful approach is again to organise the raw data into a frequency distribution, which we can then chart if necessary. Whether to use an ungrouped or grouped frequency distribution depends on the number of different values that the variable can take:

■ If this number is relatively small (say up to 15) we can use an *ungrouped* frequency distribution. This will most often be the case with *discrete* variables.
■ With higher numbers of possible values (usually the case with *continuous* variables) we will need a *grouped* frequency distribution.

FREQUENCY DISTRIBUTIONS WITH DISCRETE DATA

We can start with the situation where the number of possible values the variable can take is small enough to use an ungrouped frequency distribution. In essence this amounts to sorting the raw sample values into ascending order and then counting how many times each value occurs in the sample data.

An Example from Practice

Table 4.1 contains the raw sample data on the number of still and live births to each woman in a sample of 138 women attending a family planning clinic and diagnosed as suffering from endometriosis[1]. These data are metric *discrete* since the number of values taken by the variable "number of births" is finite, i.e. it can take only a limited number of values. Since we wouldn't expect any women to have had say more than 15 children (or not many more anyway!) this is a small enough number for us to use an ungrouped frequency distribution*. From the raw sample data we can count the number of women who have experienced no previous births, the number who have experienced one previous birth, the number who have experienced two previous births, and so on. The maximum number in the sample data is five births, so we will need six rows in the frequency distribution. The ungrouped frequency distribution is shown in the first two columns of Table 4.2 (for the moment ignore the last column).

Table 4.1: Raw data on the number of still and live births to a sample of 138 women attending a family planning clinic and diagnosed as suffering from endometriosis

```
0 2 2 3 2 1 2 2 3 3 2 3 1 2 0 2 3 2 1 1
1 2 4 3 2 1 2 2 2 1 2 0 0 2 1 2 3 5 1 3
2 1 1 2 3 2 4 3 2 3 0 2 2 1 2 2 0 2 2 2
4 1 2 3 2 2 2 2 2 5 2 3 3 3 1 2 1 3 1 2
3 3 3 2 2 0 3 3 1 2 1 2 3 2 2 2 1 1 0 2
2 3 3 4 4 4 1 2 2 2 2 1 2 2 2 4 2 1 2 0
  2 4 2 0 2 2 3 0 2 1 3 2 2 1 2 2 2 2
```

Table 4.2: Frequency distribution for the raw data in Table 4.1

NUMBER OF BIRTHS	NUMBER OF WOMEN	PERCENTAGE OF WOMEN	CUMULATIVE PERCENTAGE OF WOMEN
0	11	8.0	8.0
1	25	18.1	26.1
2	66	47.8	73.9
3	26	18.8	92.7
4	8	5.8	98.5
5	2	1.4	100.0
Totals	138	100.0	

The frequency distribution shows for example that 11 (or 8%) of the women had not given birth before, 25 (or 18.1%) had previously had one child, and so on. The largest value taken by the variable is five children (only two women, or 1.4% of those in the sample). It is obviously much easier to detect these features than it would have been from the raw data, even though this is a relatively small sample.

When the number of possible values that the discrete metric variable can take is larger than say 15, then we will usually be better off using a grouped frequency distribution. For example, Table 4.3 shows the grouped frequency distribution for the number of beverages a week consumed by over 13 000 male and female subjects in a study investigating the relationship between a number of demographic and other factors on alcohol intake and mortality[2].

---

* Recall that there is no *rule* about the maximum number of classes it is feasible to have in a frequency distribution before grouping is a requirement; however, a maximum 15 is a useful rule of thumb.

Table 4.3: Distribution of alcohol intake among 13 285 subjects

BEVERAGES A WEEK	NUMBER OF MEN	NUMBER OF WOMEN
< 1	625	2472
1–6	1183	3079
7–13	1825	1019
14–27	1234	543
28–41	585	72
42–69	388	29
> 69	211	20

The variable "number of beverages" is clearly discrete (it can take only a limited number of values). The table shows that there are at least 70 different values in the sample; which, if the frequency distribution was not grouped, would require a table with at least 70 rows—hardly likely to be very helpful in providing a broad view of the data. With the grouped distribution, however, we can see immediately that most male subjects consumed in the region of 7 to 13 beverages a week, and comparatively few more than 41 beverages per week. Female intake is clearly lower, most females consuming between one and six beverages a week, with comparatively few consuming more than 27 beverages a week. These broad patterns would be much more difficult to detect in the 13 285 raw sample values!

FREQUENCY DISTRIBUTIONS WITH CONTINUOUS DATA

Turning now to the continuous data case, it is more than likely that the number of possible values will be large enough to necessitate a grouped frequency distribution.

An Example from Practice

Consider the ungrouped data in Table 4.4, which records the duration (measured in seconds) of the first and fourth seizure for a group of 40 patients receiving ECT treatment for severe depression.

Even though the duration of the seizures is measured to the nearest second and has therefore the *appearance* of discrete data, it is nonetheless continuous. It is just impractical to measure it more accurately than this (although with a digital electronic clock or an ECG we might be able to measure seizure duration to several decimal places). A grouped frequency distribution using these raw data is shown in Table 4.5. This distribution has been designed with ten classes, each with a width of 10 seconds.

The organisation of the raw data into a grouped frequency distribution gives us a much clearer view of the broad pattern of seizure duration than scanning the raw data would have done (and this advantage would be even more pronounced with larger samples). Table 4.5 shows immediately, for example, that the most common seizure duration for the first seizure is between 30 and 40 seconds, and this is also true of the fourth seizure. (Note however that there is a noticeable difference in the spread of seizure durations between the two situations, a point we will return to in the next chapter.)

Table 4.4: Raw data for the duration of seizure of patients having first and fourth treatments with ECT

PATIENT NUMBER	DURATION OF SEIZURE (seconds)		PATIENT NUMBER	DURATION OF SEIZURE (seconds)	
	First	Fourth		First	Fourth
1	45	40	21	90	50
2	37	40	22	30	20
3	30	21	23	15	40
4	40	20	24	40	42
5	60	45	25	42	50
6	30	17	26	30	10
7	45	28	27	20	30
8	30	35	28	35	35
9	30	35	29	25	25
10	35	35	30	25	30
11	64	30	31	15	10
12	32	21	32	30	40
13	25	30	33	30	20
14	50	30	34	45	25
15	33	36	35	15	35
16	25	30	36	45	20
17	50	25	37	40	40
18	25	30	38	30	25
19	20	30	39	15	20
20	50	40	40	30	25

Table 4.5: Grouped frequency distributions for the duration of the first and fourth ECT seizures

DURATION (seconds)	FIRST	FOURTH
0–9	0	0
10–19	4	3
20–29	7	13
30–39	15	14
40–49	8	8
50–59	3	2
60–69	2	0
70–79	0	0
80–89	0	0
90–99	1	0

It so happens that Table 4.5 has been designed with all classes the same width. Where possible this is a good practice to follow because if, say, one or more of the class widths is different (e.g. the third class might have been 30–49), this might not be noticed by somebody examining the frequency distribution and thus be misleading. Also, having different class widths has implications for charting the distribution (we will return to this point shortly).

OPEN-ENDED CLASSES

There is one modification to a frequency table (either grouped or ungrouped) which is sometimes very convenient. This is the introduction of an *open-ended class*, which is often encountered in reports and journal articles. We can use open-ended classes with discrete as well as continuous metric data (and indeed with numeric ordinal data if we wish). To illustrate the idea, notice that the last three classes in Table 4.5 only have a frequency of 1 between them for the first seizure, and none at all for the fourth seizure. Some of these classes therefore seem redundant. In such circumstances classes are often amalgamated into a single open-ended class.

"It was definitely a number 2, officer."

Doing this with Table 4.4, by aggregating the last three classes into a single "70 & over" class, produces the frequency distribution of Table 4.6. In other words no exact upper value for the top class is specified. If we have access to the raw data we will of course know what the actual upper value should be, but if we are presented with an open-ended class in a published paper or report and do not have access to the raw data, we will be unable to determine what this value is, although we might be able to make an intelligent guess. As a presentation device, therefore, open-ended classes have the convenience of being more compact but suffer the disadvantage of not providing quite as much information. Open-ended classes can be used at either end or both ends of the frequency distribution.

Table 4.6: Open-ended frequency distributions for first and fourth ECT seizures

SEIZURE DURATION (seconds)	FIRST SEIZURE	FOURTH SEIZURE
0–9	0	0
10–19	4	3
20–29	7	13
30–39	15	14
40–49	8	8
50–59	3	2
60–69	2	0
70 & over	1	0

CUMULATIVE FREQUENCY DISTRIBUTIONS

We first encountered the idea of cumulative frequency in the previous chapter. Its use with quantitative data is equally straightforward. For example, for the ECT data, suppose we ask the question "How many (or what percentage of) seizures lasted less than 30 seconds?, or less than 50 seconds?". We can best answer questions of this sort by calculating *cumulative frequency*.

As we saw in Chapter 3, the procedure for calculating cumulative frequency is very simple—we simply add up the values in the frequency or relative frequency column, one by one, starting at the top of the column. Table 4.7 shows the values of cumulative frequency and relative cumulative frequency for the first ECT seizure duration using Table 4.5.

Table 4.7: Cumulative and relative cumulative frequency distributions for the duration of the first ECT seizure

DURATION (seconds)	FREQUENCY (Number of Seizures)	RELATIVE FREQUENCY (% of Seizures)	CUMULATIVE FREQUENCY	RELATIVE (%) CUMULATIVE FREQUENCY
0–9	0	0.0	0	0.0
10–19	4	10.0	4	10.0
20–29	7	17.5	11	27.5
30–39	15	37.5	26	65.0
40–49	8	20.0	34	85.0
50–59	3	7.5	37	92.5
60–69	2	0	39	97.5
70–79	0	0	39	97.5
80–89	0	0	39	97.5
90–99	1	0	40	100.0

Now we can see at a glance, by reading across the appropriate row of the table, that 11 seizures (or 27.5%) lasted less than 30 seconds, or 34 (85.0%) lasted less than 50 seconds. Notice that we cannot use the cumulative frequency distribution of Table 4.7 to answer the question "How many (or what percentage of) seizures lasted less than 25 seconds? or 48 seconds?", but only questions which correspond to one of the class limit values. As we shall see shortly there is a graphical procedure for doing just this.

Cumulative frequencies can be calculated for metric variables that are discrete as well as those that are continuous. For example, from the last column of Table 4.2 we see that 73.9% of women had had two or fewer children.

CHARTING CONTINUOUS QUANTITATIVE DATA

We saw in Chapter 3 that bar and pie charts are commonly used to chart *qualitative* discrete data. With *quantitative* data, the choice of an appropriate chart depends on whether the data are continuous or discrete. Let us consider the case of *continuous* data first. By far the most popular charting method is the *frequency histogram* used to display a grouped frequency distribution.

The Frequency Histogram

Figure 4.1 is a frequency histogram for the grouped frequency distribution of the first ECT seizure durations in Table 4.5. Although similar in appearance to a bar chart (frequency on the vertical axis, values of the variable on the horizontal), the histogram differs in several significant respects:

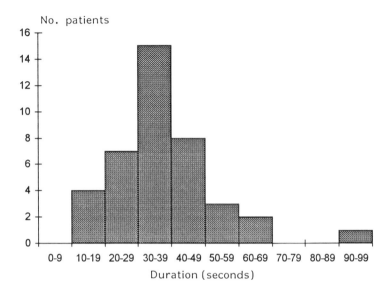

Figure 4.1:   Histogram of duration of first ECT seizure (seconds) from data in Table 4.5

- First of all it has *no* spaces between the bars, to reflect the fact that we are dealing now with continuous data rather than discrete categories.
- Second, if all the classes in the grouped distribution have the same width, then the height of each bar in the histogram is equal to its frequency, as it is in the bar chart.
- If the classes have different widths, then the heights of the histogram bar(s) have to be adjusted accordingly.

As an example of this last, and very important, point, if one class has a width twice that of the others, the height of the corresponding bar in the histogram should be halved, and vice versa; if three times as wide the height should be divided by three, and vice versa. (Note that these adjusted frequencies are known as frequency *densities*.) The effect of these adjustments is to maintain the "area proportional to frequency" property of the histogram, which is important for a correct interpretation.

Finally, we cannot construct histograms for distributions with open-ended classes (such as Table 4.6), since we have no lower and/or upper class value(s) for the open-ended class(es), and thus cannot complete the histogram. We may be able to get round this problem by making some reasonable assumption about the value of the lower or upper limit of an open-ended class. With Table 4.6 for example, we might assume that no seizure will last more than 99 seconds, and assume an upper limit for the open-ended class of 99 seconds, i.e. have a top class of (70–99)

seconds. (The width of this class would then be three times wider than the other classes and its frequency would thus have to be divided by three before drawing the histogram.)

The Frequency Polygon

An alternative graphical presentation of continuous quantitative data is the *frequency polygon*. Instead of bars, points equal in height to frequency are placed over the appropriate *class midpoints*. These points are then joined with straight lines. It is necessary to close the polygon at each end by assuming an imaginary class with zero frequency at each end of the distribution. The class midpoints for the frequency distribution of the duration of the first ECT seizure (Table 4.5) are 4.5, 14.5, 24.5, and so on. The frequency polygon for the first ECT seizure duration is shown in Figure 4.2.

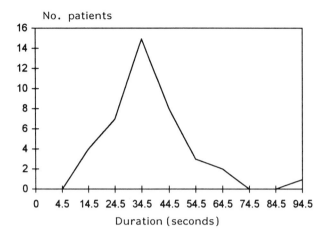

Figure 4.2: Frequency polygon for duration of first ECT seizure

One advantage of the polygon over the histogram is that it is easier to compare the shapes of two frequency distributions using two superimposed polygons rather than two superimposed histograms. This idea is illustrated in Figure 4.3, which shows the frequency polygons for the duration of the first and fourth ECT seizures.

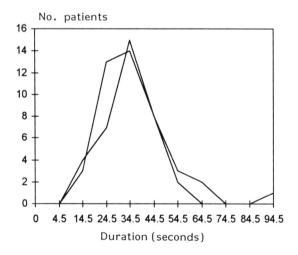

Figure 4.3: Frequency polygons for first and fourth ECT seizure durations

The Cumulative Frequency Curve or Ogive

We have already seen an example of an ogive in Chapter 3. The ogive for the relative cumulative frequency distribution of Table 4.7 (first ECT seizure duration) is shown in Figure 4.4. We can use this curve to determine the value of the percentage of seizures which lasted less than any value in seconds we wish simply by reading the chosen value on the horizontal axis, moving vertically to the curve and then horizontally to the percentage axis.

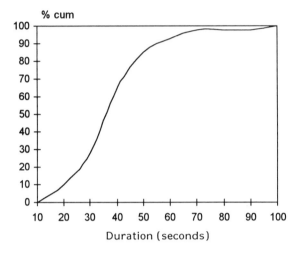

Figure 4.4: Cumulative frequency curve or ogive for duration of first ECT seizure

For example, the number of seizures lasting less than 25 seconds can be read as 20% (approximately). Alternatively we can reverse this process and estimate the length of time that any given number or percentage lasted less than. For example, 50% of seizures lasted less than about 36 seconds. Many statistics packages will provide cumulative frequency curves.

Using a Computer to Chart Data

All of the charts in this chapter, with the exception of the dot plot, were produced using the Microsoft Graph facility in Word for Windows 6.0. The procedure is very straightforward, and I described it in Chapter 3, for both Word and Excel (review if necessary). However, it's worth mentioning that to produce a histogram from a column chart, the *Gap Width* has to be set to 0. If Word or Excel is not available, then DrawPerfect or another spreadsheet program, e.g. dBase or Lotus 1–2–3, could be used. Alternatively, most of the dedicated statistics packages, SPSS, Minitab and UNISTAT, for example, can also be used to draw charts.

We looked at the use of SPSS for charting in Chapter 3. Briefly, the data to be charted should be entered into a column (or columns) of the datasheet, then **Graphs** clicked on with the mouse. The drop-down menu will then offer a choice of a number of different graphs including: **Bar, Line, Area, Pie,** and **Histogram**. You should select the column(s) of data you want to chart and the required graph type and then respond to the queries in the dialogue box. Alternatively, **Statistics, Summarize, Frequencies, Charts . . .,** offers the choice of bar charts or histograms, along with a frequency distribution.

In Minitab, after the data have been entered into the worksheet, the mouse is used to click on **Graph**. The drop-down menu offers a choice of graphs. The required graph should then be selected and the queries in the dialogue box responded to. In both packages the procedures are quite straightforward, although some experimentation in the beginning will be required.

CHARTING DISCRETE METRIC DATA

The Dot Plot

A simple but useful method for displaying *ungrouped* discrete data is offered by the Minitab package and known as the *dot plot*, in which the vertical axis measures frequency, and the values of the variable are marked on the horizontal axis. One dot for every value in the sample is placed above its appropriate value on the horizontal axis. On occasion the dot plot can also be used to display *continuous* metric data if the number of possible values is not too large (as will often be the case if the variable values are integers, i.e. whole numbers) and/or the sample size is big enough to provide a sufficient number of multiple values. The dot plot may also be found helpful when charting numeric *ordinal* data. A dot plot is available via **Character Graphs** in the Minitab **Graph** menu. Figure 4.5 is a dot plot of the duration of the first ECT seizure using the ungrouped data in Table 4.3 (and produced by Minitab).

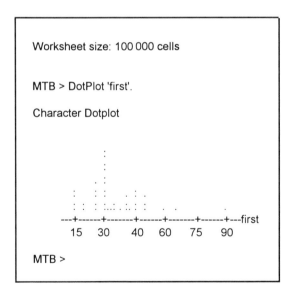

Figure 4.5:   Dot plot of the duration of the first ECT seizure, using the ungrouped data in Table 4.3 (and produced by Minitab)

The dot plot gives an immediate impression of the distribution of individual ECT seizure durations. We see, for example, that most of the seizures lasted somewhere between 20 and 40 seconds, and that more lasted 30 seconds than any other time.

The Frequency Diagram

As an alternative to the dot plot, we can display an *ungrouped* frequency distribution for *discrete* metric data, such as that for the number of previous births by the endometriosis women in Table 4.1, using a *frequency diagram*. The idea is illustrated in Figure 4.6. Rather than using bars to illustrate each class frequency we use vertical lines centred on each class value.

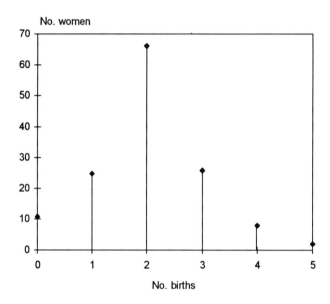

Figure 4.6: Use of a frequency diagram to show the number of previous births experienced by women diagnosed as having endometriosis

The Step Chart

If we want to display an ungrouped *cumulative* frequency distribution for a discrete metric variable, we can do so using a step chart, an example of which we have already seen in Chapter 3.

The Frequency Curve

The reason for constructing frequency distributions, histograms and ogives, is to get a clearer understanding of what patterns exist in the sample data. If we draw a smooth curve through a frequency histogram or through a polygon we get the ultimate realisation of this pattern. This curve is known as the *frequency curve*. Figure 4.7 shows the frequency curve superimposed on the polygon for the first ECT seizure.

Skewed Frequency Curves

The frequency curve in Figure 4.7 is not quite *symmetrical*. The tail on the right extends a little further than that on the left. Such departures from symmetry in a frequency distribution are known as *skewness*. A frequency distribution may be either left- or right-skewed (also known as negative and positive skew). The frequency curve in Figure 4.7 is (slightly) positively- or right-skewed. Figures 4.8, 4.9 and 4.10 illustrate examples of left-, symmetric and right-skewed frequency curves.

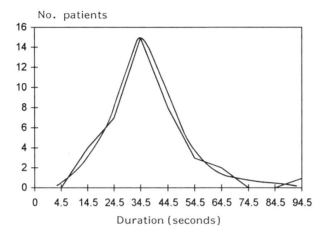

Figure 4.7: Smooth frequency curve drawn through a frequency polygon for duration of first ECT seizure

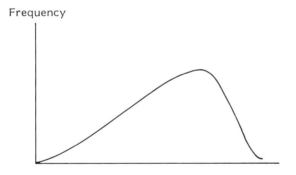

Figure 4.8: A left or negatively skewed distribution

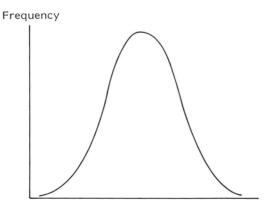

Figure 4.9: A symmetric distribution

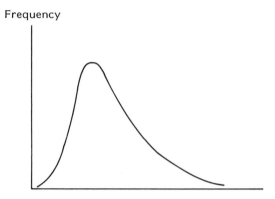

Figure 4.10:   A right or positively skewed distribution

THE NORMAL DISTRIBUTION

There is one special and important shape that a symmetrical frequency curve can sometimes take which is known as a *Normal distribution*. A "Normal" frequency distribution has a smooth symmetric bell-shaped curve that can be described by a mathematical equation (the nature of which does not need to concern us in this book). The idea of a "Normal" frequency distribution is extremely important in statistics. Remember that this is special use of the word "Normal" (i.e. it does *not* mean "usual"). Figure 4.11 is an example of a Normal frequency curve.

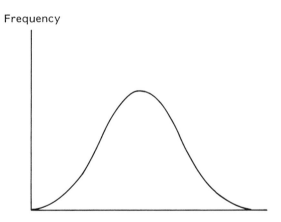

Figure 4.11:   The "Normal" frequency curve

As you will see, both SPSS and Minitab produce a numeric measure of the degree of skewness of any distribution. Values of the skewness coefficient equal to 0 indicate a symmetric distribution, although with sample data such a value is very unlikely, even if the population is symmetrically distributed. Positive values of the skewness coefficient indicate (not surprisingly) positive skew, negative values indicate negative skew. The further the value is away from 0 the more skewed the distribution is.

The shapes of frequency distributions are also described by their degree of *kurtosis*. Distributions whose curves are *flatter* than the Normal distribution are described as being *platokurtic*, distributions "peakier" or more "pointy" than the Normal distribution are described as being *leptokurtic*. Both SPSS and Minitab calculate values for the coefficient of kurtosis. A zero value for the coefficient of kurtosis indicates a Normal distribution; negative values a peaky (leptokurtic) distribution, positive values a flatter (platokurtic) distribution. The further away from 0, the bigger the departure from "Normality".

It can be very helpful in judging the "Normal-ness" of a frequency distribution if a Normal curve is superimposed on top of a frequency histogram for the distribution in question, and both SPSS and Minitab allow this option. As an example, we can superimpose a Normal curve on the histogram for the duration of the first ECT seizure in Figure 4.1. Assume that the data are in column 1 and labelled "First"; then in SPSS the commands:

   Graph
     Histogram
       Select First
         ☒ **Display Normal Curve**
        OK

will produce the graph shown in Figure 4.12. The procedure in Minitab is much more awkward and I will leave it to you to explore the Minitab reference manual if you are interested.

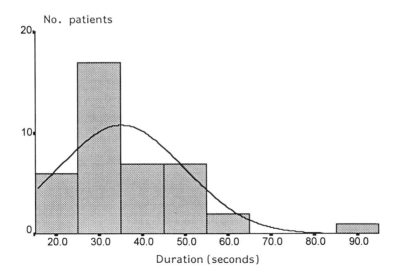

Figure 4.12: The Normal curve superimposed on the frequency histogram for the duration of the first ECT seizure

SUMMARY

In this chapter we have looked at ways of organising and presenting quantitative data, i.e. data generated by the measurement of metric variables, both discrete and continuous.

As with qualitative data, the frequency (and cumulative frequency) distributions remain the most useful tabular methods for organising data so that any *broad* underlying patterns and characteristics are quickly revealed.

If we wish to chart the data, we have a wider choice of approaches than with qualitative data. The frequency histogram and frequency polygon are the most widely used methods for charting continuous data, with the cumulative frequency curve (or ogive) used to chart cumulative frequency.

For discrete data we can use the dot plot, and for charting cumulative discrete data, the step chart is the most appropriate device.

Frequency distributions (and their corresponding frequency curves) may be symmetric, left skewed, or right skewed. The "Normal" distribution is a special symmetric frequency distribution which when plotted has a smooth bell-shaped curve. As we will see later, the Normal distribution is very important in many statistical applications.

EXERCISES

4.1 In a study into the relationship between passive smoking and heart disease, researchers counted the number of smokers among the co-workers of each subject in the study[3]. The adapted data in Table 4.8 are from a sample 45 individuals and record the number of smokers among their co-workers.

   (a) What kind of data are these?
   (b) Why would you choose an ungrouped rather than a grouped frequency distribution when organising these data?
   (c) Construct the ungrouped frequency distribution for the data. What broad patterns are thus revealed?
   (d) Construct the frequency diagram from the frequency distribution calculated in (c). Does this offer any additional insights over the frequency distribution?
   (e) Calculate the cumulative frequency distribution. (i) How many subjects had fewer than five smoking co-workers? (ii) A quarter of subjects had fewer than how many smoking co-workers?
   (f) Chart the cumulative frequency distribution using a step graph. Does this offer any further insights into the sample data?
   (g) Explain why you might introduce an open-ended class into this frequency distribution.

4.2 The times (in minutes) spent waiting to see a GP by a sample of 60 patients at a large health centre are shown in Table 4.9.

   (a) Construct a grouped frequency distribution for waiting time. Comment on what is revealed about the distribution of waiting times.

Table 4.8:   Number of smoking co-workers for a sample of 45 non-smoking females

SUBJECT	1	2	3	4	5	6	7	8	9	10	11	12	13	14	15
SMOKERS	2	0	5	15	1	8	5	0	0	9	10	4	4	2	3
SUBJECT	16	17	18	19	20	21	22	23	24	25	26	27	28	29	30
SMOKERS	5	1	14	7	9	2	3	0	0	5	8	1	6	0	6
SUBJECT	31	32	33	34	35	36	37	38	39	40	41	42	43	44	45
SMOKERS	0	4	7	5	2	2	6	3	1	0	1	1	3	2	0

(b) Draw the corresponding histogram and polygon. Comment on the relative merits of the frequency distribution and the charts.
(c) Calculate the relative cumulative frequency distribution. What percentage of patients wait (i) less than 10 minutes; (ii) less than 20 minutes. A quarter of patients wait less than how many minutes?

Table 4.9:   Waiting times (in minutes) of 60 patients of a GP practice

PATIENT	1	2	3	4	5	6	7	8	9	10	11	12	13	14	15
WAITING TIME	2	34	14	15	14	17	12	10	8	26	21	15	20	5	26
PATIENT	16	17	18	19	20	21	22	23	24	25	26	27	28	29	30
WAITING TIME	11	12	18	9	22	21	15	12	8	10	29	24	16	16	19
PATIENT	31	32	33	34	35	36	37	38	39	40	41	42	43	44	45
WAITING TIME	13	8	8	30	17	12	10	32	18	22	4	15	5	13	19
PATIENT	46	47	48	49	50	51	52	53	54	55	56	57	58	59	60
WAITING TIME	6	27	7	12	12	25	14	18	15	24	20	16	14	18	10

4.3 In a study of the relationship between low iron status and psychological functioning[4], researchers measured the haemoglobin levels of a sample of 130 males and 165 females. The grouped frequency haemoglobin distributions are shown in Table 4.10.

(a) Comment on the way in which the authors have designed the grouped frequency distributions.
(b) What is revealed about the distributions of male and female haemoglobin levels by the frequency distributions?
(c) Make some reasonable assumptions about the open-ended class values and hence draw the histograms for males and females.
(d) Draw the two frequency polygons on the same graph and comment on any male–female differences in haemoglobin levels.
(e) Construct cumulative *relative* frequency distributions for males and females and draw the corresponding ogives. What percentage of males and females respectively have haemoglobin levels of (i) 15 g/dl or less; (ii) less than 14.5 g/dl. What haemoglobin levels do a half of the males and females respectively have less than?

Table 4.10:   Haemoglobin levels in samples of males and females

HAEMOGLOBIN (g/dl)	MALES	FEMALES
< 12	0	11
12.1–13	2	47
13.1–14	8	76
14.1–15	52	27
15.1–16	45	4
> 16	23	2

REFERENCES

1. Vessey, M. P. *et al.* (1993) Epidemiology of endometriosis in women attending family planning clinics. *BMJ,* **306**, 182–4.
2. Groenbaek, M. *et al.* (1994) Influence of sex, age, body mass index and smoking on alcohol intake and mortality. *BMJ,* **308**, 302–6.
3. He, Y. *et al.* (1994) Passive smoking at work as a risk factor for coronary heart disease in Chinese women who have never smoked. *BMJ,* **308**, 380–9.
4. Fordy, J. and Benton, D. (1994) Does low iron status influence psychological functioning? *Journal of Human Nutrition and Dietetics,* **7**, 127–33.

5

MEASURES OF AVERAGE

❑ SUMMARY MEASURES OF AVERAGE ❑ FACTORS GOVERNING THE CHOICE OF A SUITABLE MEASURE OF AVERAGE ❑ THE MODE, MEDIAN AND MEAN FOR QUALITATIVE AND QUANTITATIVE DATA ❑ QUALITIES OF EACH MEASURE ❑ APPROPRIATENESS OF EACH MEASURE ❑ COMPARING MEASURES ❑ OTHER MEASURES OF AVERAGE, CENTILES ❑

GETTING DOWN TO DETAILS

We have seen in the previous two chapters that one of the best ways of getting a quick overall view of the sample data is to construct a frequency distribution, and maybe also draw some sort of chart. This is okay as far as it goes, but we will often want to be a bit more precise than this, and put some finer detail into our description of the data. We can do this by calculating what are known as *summary measures*. In descriptive statistics, two summary measures are widely used. The first of these tells us what the average of the sample values is, the second tells us how spread out the sample values are. Collectively these measures are known (not surprisingly) as measures of *average** and measures of *spread*. This chapter concentrates on measures of average, the next on measures of spread.

Spoilt for Choice?

There are three commonly used measures of average to choose from:

- the *mode*
- the *median*
- the *mean*.

Each one of these measures interprets the idea of "average" in a different way, and our choice of an appropriate measure for any given set of sample data will depend on two factors:

- what *aspect* of "averageness" we want to capture
- what kind of data we are dealing with—nominal, ordinal or metric.

---

* Measures of average are also known as measures of location and measures of central tendency.

The second consideration is crucial, because we don't always have a completely free choice from among the three measures. The usual situation you will face in practice is that you will have a set of sample data of some particular type—it may be nominal (perhaps different types of drug prescribed by a doctor); it may be ordinal (for example a set of Waterlow scores); it may be metric (perhaps waiting times in an A&E department). Whatever the circumstances, you want to calculate the average of the sample values and you need to decide which is the most appropriate measure of average. The first thing to consider is the type of variable involved.

Measures of Average and Type of Variable

In Chapter 1 we spent quite some time examining the differences between nominal, ordinal and metric variables and their data. This discussion was important because the choice of a measure of average is *crucially* dependent on the type of variable in question, as follows:

- If the variable is *metric* we have a completely free hand and can choose either the mode, the median or the mean as a measure of average. The choice will be dictated principally by what aspect of average we wish to capture.
- If the variable is *ordinal* the mean should not be used, and the choice is therefore between the median and the mode.
- If the variable is *nominal* then the only measure of average is the mode.

Table 5.1 summarises these choices. With these considerations in mind, the remainder of this chapter discusses measures of average for metric data, ordinal data and nominal data in turn. We will see how each measure is calculated, what precisely is being measured, and the advantages and disadvantages of each.

Table 5.1:   Choice of measure of average and type of variable

MEASURE OF AVERAGE	TYPE OF VARIABLE		
	Nominal	*Ordinal*	*Metric*
Mode	yes	yes	yes
Median	no	yes	yes
Mean	no	no	yes

MEASURES OF AVERAGE FOR METRIC DATA

If the data in question are metric, all three measures of average can be used as a summary measure of the sample data. The choice between them depends on what aspect of average you wish to capture and the respective properties of each measure.

The Mode for Metric Data

The mode interprets the meaning of *average* as the value which occurs *most often* in the data. The mode is thus a measure of *typicality*. If we want to answer the question "What is the most typical (or commonest) value among the sample values?" then the mode is the most appropriate measure of average. We calculate the value of the mode as follows:

- Arrange the raw sample data in ascending order.
- Count the number of times each sample value occurs.
- Identify the value (or values, if there is more than one) which occurs most often. This value is the mode.

For example, the raw data in Table 5.2(a) show the time spent in a hospital (in days) by 11 patients having a hip replacement operation. To find the mode for the hospital data we first rearrange the data into ascending order (Table 5.2(b)) and then count the number of times each value occurs. We see that the mode is 8 days (since this value occurs three times which is more than any other value). So if we want a *summary* measure of the length of stay which reflects the *typical* experience of patients we would choose the modal value of 8 days.

Table 5.2: (a) Duration of stay (days) in hospital for patients having hip replacement operations. (b) Values arranged in ascending order

(a) RAW DATA										
Stay (days):	17	8	19	8	12	15	25	5	8	23	15
(b) DATA IN ASCENDING ORDER										
Stay (days):	5	8	8	8	12	15	15	17	19	23	25

Although the mode is very easy to calculate when the raw or ungrouped data are available, its calculation from grouped data is awkward. Hence the mode is not widely used as a measure of average, suffering as it does from this and a number of other limitations.

MEMO

The mode is the value that occurs most frequently in the sample data. The mode measures the most *typical* value. There may be more than one mode.

Properties of the Mode

- The mode is not particularly suitable as a measure of average for *continuous* metric variables, since the number of different values in the sample data will usually be large (especially if the data are decimal). Indeed no value may occur more than once, in which case we would have as many modes as there are different values.
- Calculation of the mode does not use all of the information in the sample, i.e. it does not involve all of the sample values, but only those that occur the most often. This could be considered a serious weakness if we feel that all of the available information in the sample should be used.
- Because of the above, the mode may not actually be very descriptive of the typical experience of the sample. For example, 7 out of the 11 patients in the above example had a stay longer than the mode of 8 days.
- The mode may not be unique, i.e. there may be more than one mode. For example, if the patient who stayed 12 days had stayed 3 days longer we would have had two modes, of 8 and 15 (both occurring three times). Hardly a helpful description of typical stay times! Note that distributions with two modes are referred to as *bimodal*.
- We cannot combine modes to find the average mode of two separate distributions (say the modal stay times for two hospitals combined).
- The mode is a *volatile* measure, this means that it is sensitive to quite small changes in the sample values.

On the other hand, the mode has a number of attractive features:

- It is not particularly affected by extreme values, known in statistics as *outliers*. These are values that are much smaller or much larger than the general mass of values. The inclusion of an outlier value may result in a measure of average that is unrepresentative and therefore misleading.

- It is an easy concept to understand and calculate (at least with ungrouped data).
- The value of the mode will always be equal to one of the original sample values.
- It can be found graphically.

But as a consequence of its limitations, the mode is not widely used with metric variables where alternative measures of average are available. However, where some measure of "peak" or maximum value is required the mode is useful. An example might be planning staff numbers in an A&E or intensive care unit, where the peak number of patient arrivals (as measured by the mode) would offer a useful starting point in planning adequate clinical and medical cover.

"It's a bimodaldery, son."

This discussion has enabled us to touch upon some of the desirable qualities we might seek in any measure of average. It also provides a useful baseline against which we can compare the properties of the other two measures of average we will consider shortly.

Finding the Mode Graphically

The modal value is easily identified from a frequency diagram or dot plot charted from ungrouped *discrete* metric data. For example, the dot plot produced by Minitab for the hospital stay data in Table 5.2 is shown in Figure 5.1.

The mode is then the value under the tallest column of dots, i.e. 8 days. For *continuous* metric data that have been grouped, an *approximate* value for the mode (known as the *crude* mode) is the point on the horizontal axis directly below the midpoint of the tallest bar in the frequency histogram (or alternatively below the highest point in a frequency polygon)*.

---

* I have omitted calculations of the exact mode when only a grouped frequency distribution is available.

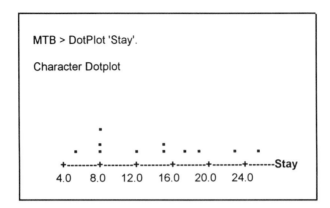

Figure 5.1: Minitab dot plot for hospital stay data in Table 5.2

The Median for Metric Data

The median identifies the *middle* value among the sample values. The median is thus a measure of *centrality*. Half of the sample values will be larger than the median value and half will be smaller*. The median is calculated as follows:

- Arrange the raw sample observations into ascending order.
- Identify the middle observation.
- The value of the middle observation is the median value.

For example, from the ascending hospital stay data in Table 5.2(b) the middle value is 15 days:

    5 8 8 8 12 *15* 15 17 19 23 25

We see that the median value leaves the same number of values (five) to the left of the median as to the right. So if we want a summary measure of average which identifies the central value we would choose the median value of 15 days.

In the above example there is an odd number of sample values, so finding the middle value is straightforward. If there is an *even* number of sample values the median is the average of the two "middle" values. For example, suppose there were only 10 patients in the hospital:

    5 8 8 8 *12* *15* 15 17 19 23

The two middle values are 12 and 15 (leaving four values below and four above). The median is the average of these two values, i.e. the median equals $(12 + 15) \div 2$ $= 27 \div 2 = 13.5$ days.

---

* Strictly speaking, half of the values will be equal to or less than the median value and half equal to or more than the median value.

MEMO

The median is the middle value after the sample values have been arranged in ascending order.

An Algebraic Procedure for the Median

Although the algebra in this section is not all that difficult, some readers may want to omit it first time round. We can express the procedure for finding the median algebraically in a manner that will come in handy later on. Suppose we have n sample values (n is conventionally used to denote sample size, so for example $n = 11$ in the hospital stay example). The procedure is then as follows:

- *Step 1:* Organise the n sample observations into ascending order and number their places in that ordering, i.e. 1st, 2nd, 3rd, etc.
- *Step 2:* Calculate the value of $(50/100) \times (n + 1)$. We use $50/100$ because we want the middle or 50% value. The $(n + 1)$ automatically corrects for whether n is odd or even.
- *Step 3:* Identify the $(50/100) \times (n + 1)$th observation; its value is the value of the median.

For example, using the stay in hospital data with 11 observations, we have:

- *Step 1:* The data are arranged in ascending order and the observations numbered in order:

Stay	5	8	8	8	12	15	15	17	19	23	25
Observation	1st	2nd	3rd	4th	5th	6th	7th	8th	9th	10th	11th

- *Step 2:* The value of $(50/100) \times (n + 1)$ is $\frac{1}{2}(n + 1) = \frac{1}{2}(11 + 1) = \frac{1}{2} \times 12 = 6$.
- *Step 3:* The value of the 6th observation is 15 days, which is the median (the same result as before).

Just to make sure the method also works when there is an even number of sample observations, let us repeat the process when there are 10 sample values:

- *Step 1:* The data are arranged in ascending order and the observations numbered in order:

Stay	5	8	8	8	12	15	15	17	19	23
Observation	1st	2nd	3rd	4th	5th	6th	7th	8th	9th	10th

- *Step 2:* The value of $\frac{1}{2}(n + 1)$ is $\frac{1}{2}(10 + 1) = \frac{1}{2} \times 11 = 5.5$.
- *Step 3:* The median is the value of the 5.5th observation, i.e. half-way between the values of the 5th and 6th observations, which is the average of 12 and 15, or 13.5 days (the same result as before).

MEMO

Algebraically the median is the value of the $\frac{1}{2}(n + 1)$th observation, where n is the number of observations in the sample arranged in ascending order.

Properties of the Median

- The median is not as volatile as the mode, i.e. it is reasonably stable if small changes are made to the data.
- The median, like the mode, is not particularly sensitive to extreme values (outliers).
- The median can only ever take one value (unlike, as we have seen, the mode, which can be bimodal, or worse).
- The median will only be equal to one of the original sample values if the sample size is odd.
- Like the mode, the median does not use all of the information present in the sample, since the numeric value of the median is calculated using only the middle value (or two middle values if the sample size is even); the rest of the sample data is disregarded.
- The median value is closer to all the other values in the sample than any other measure of average would be.
- We can't combine two or more medians to find their overall average value.
- The median can be determined graphically.

Determining the Median Graphically

It is possible to determine the median graphically from a frequency distribution using either the cumulative frequency curve (or ogive) in the case of a frequency distribution with grouped classes, or the step diagram with an ungrouped frequency distribution. As an example, consider the ogive for the duration of the first ECT seizure shown in Figure 4.4. For convenience this is reproduced in Figure 5.2.

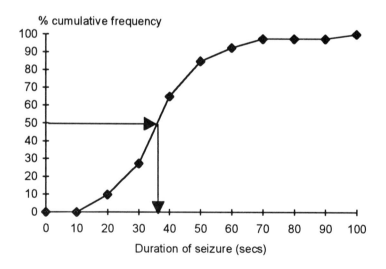

Figure 5.2:   Finding the median graphically from the cumulative percentage curve (ogive) for duration of first ECT seizure (data from Table 4.7)

Since the vertical axis is marked in percentage terms, we locate the 50% value (or the value of $(n + 1)/2$th observation if the frequency is not in percentage terms), draw a line from this point horizontally to the ogive, and a line from the point of intersection down to the horizontal axis, where we can read the median value as approximately 36 seconds.

The Arithmetic Mean for Metric Data

The arithmetic mean is none other than the common-or-garden *average* with which you will already be very familiar. As you will know, to calculate the mean, we simply add up a set of numbers and divide by however many numbers we added up. In the hospital stay example, to find the mean stay we add up the 11 stay values and divide the total by 11. This gives a value for the mean stay of $155 \div 11 = 14.1$ days.

So if we want a summary measure which captures the conventional idea of an average, we would choose the mean value of 14.1 days. Compared with the mode and median, the mean has the following properties.

Properties of the Arithmetic Mean

- Unlike with the median and the mode, calculation of the mean involves all the values in the sample, i.e. it uses *all* of the information available.
- The above property means that its value is necessarily affected by outliers.
- The mean is stable in the face of small changes to the sample data.
- The mean will not generally be equal to one of the original sample values.
- The mean cannot be determined graphically.
- The mean possesses a number of very important mathematical properties, which need not concern us here, but which make the mean the most important measure of average for metric data, and a basis for many advanced concepts in statistics. One property worth mentioning is that we can combine the means from different samples to get an overall mean of means, an operation we saw cannot be performed with either modes or medians.

MEMO

The mean is simply the conventional average. To find the mean we add up the sample values and divide by the number of values.

The Mean, Median and Mode Compared

We can now compare the values we obtained for the three measures of average for the hip-operation hospital stay time (Table 5.3). The values of the three measures of average are not all the same since, as we now know, each measure captures a different aspect of average. The only exception is when the frequency distribution for the sample data is perfectly *symmetric*, i.e. the left half of the frequency curve is the mirror image of the right half. In this case the three measures are identical (the "Normal" distribution mentioned in Chapter 4 is one such symmetric distribution).

Table 5.3: Measures of average for the time spent in hospital by patients having a hip replacement operation (raw data in Table 5.2)

MEASURE OF AVERAGE	STAY (days)
Mean	14.1
Median	15
Mode	8

To reiterate, if we want to convey the *typical*, most common, in-hospital time we would choose to use the mode; if we want to give an idea of the *central* value, which has half the values in the sample above it and half below, we would choose the median; if we want to convey the *average* value we would choose the mean. With the metric hospital data above we were able to calculate all three measures of average; however, as we will see shortly, we don't always have the luxury of a completely free choice.

MEMO

- Neither the mode nor the median use all the information in the sample.
- The mean is sensitive to the presence of outliers (extreme values); the median and the mode are not.
- The mean and the median are not affected by small changes in the sample data; the mode may be.
- The mean and the median can only take one value for a given set of sample data, the mode may have more than one value.
- Only the mode is guaranteed to have a value equal to one of the original sample values.
- The mode and the median can be found graphically.

The Summation Notation for Sample Mean

We can express the procedure for finding the mean algebraically, but to do this we need first to introduce the idea of the *summation operator*. Although the idea of the summation operator is important, you don't need to know anything about it to calculate the mean (as we have just seen), so you may prefer to omit this section at first reading. This algebraic way of expressing formulae and equations is widely employed in books and articles containing statistical material, and in this respect a familiarity with it is useful in understanding the literature.

Suppose, sticking to our example of length of stay in hospital for a hip operation, we let the variable X equal the length of stay (in days), so X_1 denotes the stay of the first patient, X_2 the stay of the second patient, and so on up to X_n which is the stay of the nth (i.e. the 11th) patient (as usual we denote the number of patients in the sample as n). The sample mean is conventionally denoted as $\bar{x}$ (pronounced "ex bar").

We have seen that to calculate the mean we add the n sample values and divide this total by n. This procedure can be written as:

$$\bar{x} = \frac{X_1 + X_2 + X_3 + \ldots + X_{11}}{n}$$

But we can express algebraic expressions like this much more succinctly if we use the *summation operator* Σ (Greek capital "sigma"). This is an algebraic operator just like, say, the sign "÷" is. When we see 8 ÷ 4 we don't need to be told that this means "divide" because it is already so familiar to us. In the same way the sign Σ tells us to "sum" a set of values. For example, when we see ΣX it means sum (i.e. add together) all the values taken by the variable X.

This allows us to express the sum of the values of X in the numerator of the above expression as ΣX. So the expression for sample mean can now be abbreviated to:

$$\bar{x} = \frac{\Sigma X}{n}$$

To repeat, this notation is widespread in statistics, but you don't need to know it simply to calculate the mean of a set of sample values.

Using the Median when Metric Data are Skewed

As a general rule of thumb, if we are working with metric data the mean is widely accepted as the most appropriate measure of average*. However, the median is often more appropriate when the data are *skewed* and/or contain *outliers*. In these circumstances it might be felt that the mean gives a misleadingly too low or too high a value (depending on the direction of the skew). The reason for this is because the mean includes *every* sample value in its calculation, and the presence of outliers, either to the left or to the right, will *drag* the value of the mean in that direction. For example, Table 5.4(a) shows the survival times (in weeks) for a sample of 24 patients from time of diagnosis of full-blown AIDS to death.

Table 5.4:   (a) Raw data on the survival times (weeks) for a sample of 24 patients dying from AIDS, and (b) arranged in ascending order

---
(a) RAW DATA
45 69 73 51 84 70 68 57 52 49 80 95 76 65 56 50 66 58 62 54 58 60 104 158

(b) DATA IN ASCENDING ORDER
45 49 50 51 52 54 56 57 58 58 60 *62* *65* 66 68 69 70 73 76 80 84 95 104 158

The data show a positively skewed distribution with the bulk of values lying between 50 and 80 weeks. There is also an outlier at 158 weeks representing one patient who survived much longer than everybody else in the sample.

If the data are sorted into ascending order, the modal survival time is seen to be 58 weeks, the most common value (occurring twice). With $n = 24$, the median

---

*We don't need at this stage to consider the reasons for this, but perhaps the principal reason lies in its mathematical properties which make it a superior estimator when it comes to statistical inference.

survival time is half-way between the value of the 12th and 13th observations*, i.e. half-way between 62 and 65 weeks. Thus the median is 63.5 weeks. The mean survival time is equal to the sum of the 24 survival times divided by 24:

$$\bar{x} = \frac{\Sigma X}{n} = \frac{1660}{24} = 69.2 \text{ weeks}$$

The mean of 69.2 weeks is considerably higher than the median of 63.5 weeks because of the effect of both the outlier value and the longish tail of values stretching to the right. In this and similar situations many of us would question whether the mean gives a proper *feel* for the *average* experience of these patients, even though it does take account of all the data. It is for this reason that the median is often used with skewed metric data (and in my opinion should be used a lot more often).

An Example from Practice

The data in Table 5.5 show measurements on sex, age and height of 10 individuals in a study into the effect of nitrogen oxide on airway response to inhaled allergens in asthmatic patients[1]. The same data are shown arranged in ascending order.

Table 5.5: Base characteristics of a sample of 12 subjects in a study of the effects of nitrogen oxide on airway responses in asthmatics

SUBJECT	SEX	AGE (years)	HEIGHT (cm)	AGE (years) (ascending)	HEIGHT (cm) (ascending)
1	F	29	160	19	155
2	F	47	167	23	160
3	F	28	165	24	165
4	M	19	178	24	165
5	F	24	155	**24**	**167**
6	M	27	165	**27**	**170**
7	M	34	182	28	172
8	M	23	173	29	173
9	F	24	170	34	178
10	F	24	172	47	182
Mean				27.9	169.0
Median				25.5	168.5
Mode				24	165

The authors calculated only the mean of the two metric variables. I have also calculated the median and mode, using the ascending-order data. In broad terms the distribution of age is positively skewed and has an outlier at 47 years. The distribution of height is less positively skewed with no outliers. The modal age is 24 years (occurring three times); the modal height is 165 cm (twice). The median age is half-way between 24 years and 27 years, i.e. 25.5 years; the median height is half-way between 167 cm and 170 cm, i.e. 168.5 cm.

---

* $\frac{1}{2}(n + 1) = \frac{1}{2}(24 + 1) = 12.5$

Compared with the median value, the mean age has a noticeably higher value, being dragged upwards by the outlier and the more pronounced skew. The mean and median heights differ less because of the more symmetric distribution of the height variable.

MEASURES OF AVERAGE WITH ORDINAL DATA

With metric data we can choose any one of the three measures of average, but we can't use normal arithmetic with ordinal data. As a consequence we shouldn't calculate the mean of a set of ordinal values. We are thus left with a choice between the median and the mode. Since this is an important point it is worth spending a little more time on it.

It is fairly obvious that we can't calculate the mean of non-numeric ordinal data. Suppose for example that we ask five patients how they feel about the service they receive when attending an outpatient department. Let us assume that the variable can take four different values and we get the following responses:

Very satisfied	1
Satisfied	3
Unsatisfied	0
Very unsatisfied	1

If we try to calculate the mean with the usual method, adding the values up and dividing by 5, we would have something like:

(very satisfied + satisfied + satisfied + satisfied + very unsatisfied) ÷ 5

which obviously we can't work out. Even if we assign numeric values to the categories (for example 0 = very unsatisfied; 1 = unsatisfied; 2 = satisfied; and 3 = very satisfied) (this operation is known as "coding"), we don't get any further forward. This is because these numeric values assume that the difference between each category is the same, i.e. that the difference between the response "satisfied" and "very satisfied" is the same as the difference between the response "very unsatisfied" and "unsatisfied". How do we know this to be true? It might be that to be very satisfied is *much* better than to be just satisfied*.

The same arguments apply to *numeric* ordinal data, such as Waterlow scores, Apgar scores, Glasgow coma scores, injury severity scores, and the like. We have already seen in Chapter 2 that the numeric values in such cases are not "proper" arithmetic values; a baby with an Apgar score of 8 is not necessarily twice as healthy as a baby with a score of 4, and the difference between scores of 2 and 3 is not necessarily the same as between, say, 5 and 6.

---

* The situation is rather analogous to the scoring system in grand prix motor racing where the difference between coming first and second might be thought to be the same as that between second and third, or between third and fourth, and so on. But in fact coming first is worth ten points, coming second only six points, coming third four points, coming fourth three points. So the "real" difference between first and second is three times the difference between second and third.

It is therefore inappropriate to calculate a *mean* score for a sample of Apgar scores, or *any* set of numeric ordinal scores. In other words, if the data are ordinal, do NOT calculate the mean as a measure of average.

This leaves us with either the median or the mode to choose from. Let us now look at each of these in turn.

The Median for Ordinal Data

Ordinal sample data may be non-numeric (such as the data in Table 3.1 on the reasons for outpatient non-attendance) or numeric (such as the Oswestry Mobility Scale data of Table 3.6). In either case, the procedure for calculating the median is exactly the same as it is for metric data described above. We first arrange the data in ascending order, then locate the position of the middle, or equivalently the $\frac{1}{2}(n + 1)$th, observation, and finally determine its value.

It is fairly unusual to want to calculate the median of non-numeric ordinal data, but as an example consider the raw ordinal data on the change in the relationships of 35 GPs following fundholding shown in Table 3.11, and suppose we're interested in the median value for the change in the relationship with consultants. Table 5.6 shows the raw data for the change with consultants arranged in alphabetic ascending order.

Table 5.6: Data from Table 3.11 on the change in the relationship with consultants after fundholding, arranged in ascending order (Key: agd = a great deal; qal = quite a lot; tse = to some extent)

OBSERVATION NUMBER	CHANGE WITH CONSULTANTS
1	agd
2	agd
3	agd
4	agd
5	agd
8	qal
9	qal
10	qal
11	qal
12	qal
13	qal
14	qal
15	qal
16	tse
17	tse
18	**tse**
19	tse
20	tse
. . .	. . .
. . .	. . .

The median change is the value of the ½(35 + 1)th = 18th observation. This 18th observation has a value of "to some extent", so the median change experienced by this sample of GPs with consultants is "to some extent".

Since it is good practice always to construct a frequency distribution when describing data, we could just as easily have calculated the median from the ungrouped frequency distribution in Table 3.12, reproduced and with the addition of a cumulative frequency column in Table 5.7.

Table 5.7: Ungrouped frequency distribution showing change in the relationship of GPs with consultants following fundholding

DEGREE OF CHANGE	NUMBER OF GPs	CUMULATIVE NUMBER OF GPs
A great deal	5	5
Quite a lot	8	13
To some extent	9	22
A little	6	28
Not at all	7	35

We seek the position of the 18th observation. There are five GPs in the first class ("A great deal"), 13 in the first two classes ("A great deal" or "Quite a lot"), which is still *five* short of the 18th GP. Since there are nine GPs in the next "To some extent" class, then the 18th observation must fall in this class. Thus the median change is "To some extent", the same result as determined before using the raw data.

If the ordinal data are numeric, but have too many different values for other than a grouped frequency distribution, then we can more easily calculate the median from the raw data, following the same steps as above. For example, the median "before" Oswestry score is found by sorting the raw data in Table 3.6 into ascending order, shown in Table 5.8.

Table 5.8: Raw Oswestry Mobility "before" scores arranged in ascending order (from raw data in Table 3.6)

OBSERVATION NUMBER	"BEFORE" SCORE	OBSERVATION NUMBER	"BEFORE" SCORE
1	10	16	70
2	10	17	72
3	10	18	73
4	10	19	80
5	10	20	80
6	10	21	90
7	15	22	90
8	25	23	95
9	25	24	100
10	35	25	100
11	43	26	100
12	55	27	110
13	60	28	110
14	60	29	112
15	63	30	115

There are 30 patients in the sample, so the median is the value of the ½(30 + 1)th observation, i.e. the value of the 15.5th observation. This is the average of the 15th and 16th values, i.e. the average of 63 and 70, which is 66.5. So the median "before" Oswestry score is 66.5 (as a matter of interest, the median "after" score is 72.5).

The Mode for Ordinal Data

Calculation of the sample mode from raw ordinal data follows the same procedure as for raw metric data; the mode is the most frequently occurring value in the sample. As an example consider again the raw non-numeric data in Table 3.11 for the change in the relationships of GPs following fundholding. We can find the mode for the change in the relationships with consultants, with other doctors in the practice, and with patients, by counting the number of times each degree of change occurs; but as indicated above, to do this in practice we would need anyway to construct the ungrouped frequency distribution shown in Table 3.12 and reproduced for convenience in Table 5.9.

Table 5.9:   The change in professional relationships of GPs as a result of fundholding

DEGREE OF CHANGE	WITH CONSULTANTS	WITH OTHER DOCTORS IN PRACTICE	WITH PATIENTS
A great deal	5	1	–
Quite a lot	8	4	1
To some extent	9	5	3
A little	6	8	5
Not at all	7	17	26
Totals	35	35	35

From Table 5.9 it is easy to see that the modes are:

>   With consultants, the mode = "to some extent" (9 occurrences)
>   With other doctors in the practice, the mode = "not at all" (17 occurrences)
>   With patients, the mode = "not at all" (26 occurrences)

If the ordinal data have a *numeric* form with a smallish number of different values, such as a sample of Apgar scores on newborn babies, we would follow the same procedure as above: first constructing an ungrouped frequency distribution and then identifying the sample value which occurs most often from this (rather than counting through the raw data).

An example of such a frequency distribution containing the Apgar scores for a sample of 98 births is shown in Table 5.10. The modal Apgar score is clearly 8, since this is the value with the highest frequency (19 births).

If the ordinal data are *numeric* with a large enough number of different sample values to require a *grouped* frequency distribution (such as the Oswestry mobility scores in Table 3.8), then we can either determine the crude mode from the grouped distribution (as described above, i.e. the midpoint of the class with the

Table 5.10: Frequency distribution of Apgar scores for a sample of 98 births

APGAR SCORE	NUMBER OF BIRTHS
0	0
1	1
2	1
3	5
4	8
5	12
6	14
7	18
8	19
9	11
10	9

highest frequency) or, if we want an exact mode, go back to the raw data and count the number of times each value occurs.

As a matter of interest, using the raw data in Table 3.6 gives a "before" mode of 10 (with six values) and "after" modes of 40 and 55 (each occurring three times, i.e. the "after" sample values are bimodal).

MEASURES OF AVERAGE FOR NOMINAL VARIABLES

As we have already noted above, the arbitrary ordering of nominal variable categories makes the use of the mean or median as a measure of average meaningless (with one exception which we will discuss in a moment), and the only feasible measure of average for nominal data is thus the mode.

An Example from Practice

Table 5.11 contains frequency distributions relating to the judgement by hospital consultants as to the appropriateness of 600 referrals by GPs and by hospital doctors (specialists)[2]. Clearly, in the opinion of the hospital consultants the mode for inappropriate referrals by GPs is orthopaedics and for referrals by specialists it is otorhinolaryngology. In other words, most *typically*, GPs make inappropriate referrals in orthopaedics, specialists in otorhinolaryngology.

The exception to finding the mean of nominal data referred to above relates to the application of the mean to *dichotomous* nominal variables. A dichotomous variable is one that can take only *two* possible values, e.g. Yes or No, Male or Female, Alive or Dead, etc. For example, consider the sample data in Table 5.5 which gives the sex of each subject in the study. If we code male = 0 and female = 1, and take the mean of the 10 values, we get the value 6 ÷ 10 = 0.6. Taking the *mean* of such dichotomous data gives the *proportion* of items coded 1 in the sample, i.e. in this case the proportion of females. Thus the value of 0.6 indicates that 60% of the sample were females.

Table 5.11: The percentage of 600 referrals by GPs and hospital doctors (specialists) judged inappropriate by hospital consultants

SPECIALITY	REFERRAL FROM GPs	REFERRAL FROM SPECIALISTS
Rheumatology	7.0	13.3
Orthopaedics	36.5	13.6
Otorhinolaryngology	1.1	20.0
Ophthalmology	9.2	7.7
Chest Medicine	0.0	0.0
Gynaecology	4.0	0.0
Totals	57.8	54.6

"I'm pretty sure it's a measure of location."

This concludes the discussion of the three most widely used measures of average. We have seen that each of these summary measures has advantages and limitations, and their use is sometimes restricted by the type of variable in question. Where a choice of measure is possible, any decision will depend on the various factors alluded to in this chapter. Note, however, that we are not necessarily restricted to reporting only *one* measure of average, and there is no reason why two or even all three measures might not be presented (variable type allowing of course). Indeed, many software packages routinely calculate all three, regardless of the variable type, and leave you the pleasure of deciding which is appropriate.

MEMO

As a rule of thumb for the most appropriate measure of average:

- Use the mean with metric data.
- Use the median with ordinal data, and consider it also with skewed metric data. Do not use the mean with ordinal data.
- Use the mode with nominal data.

SKEWNESS AND THE THREE MEASURES OF AVERAGE

In general, if the frequency curve is *left-* or *negatively*-skewed, then:

mode > median > mean

The ">" sign means "is greater than", i.e. the mode value is greater than the median which in turn is greater than the mean. This is because with a left-skewed distribution the mean is much more affected by the values in the left tail of the distribution (because the mean uses *all* of the values in the sample and is not immune to outliers), than either the median or the mode. The median value is such as to divide the total number of sample values into two halves, i.e. it divides the frequency curve into two areas of equal size. The mode value corresponds to the peak of the frequency curve. This idea is illustrated in Figure 5.3.

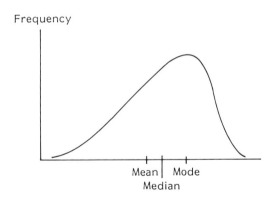

Figure 5.3: Relative positions of the mean, median and mode in a left-skewed distribution

If the frequency curve is *right-* or *positively*-skewed then:

mode < median < mean

The "<" sign means "is less than", i.e. the mode is less than the median which in turn is less than the mean. The mean is dragged upwards by the values in the right tail of the frequency curve because of its sensitivity to outliers. This situation is illustrated in Figure 5.4.

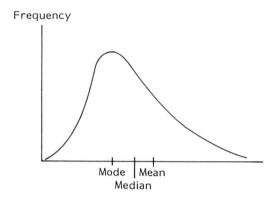

Figure 5.4: Relative positions of the mean, median and mode in a right-skewed distribution

If the frequency curve is symmetric then the values of the three measures of average will coincide. This is illustrated in Figure 5.5.

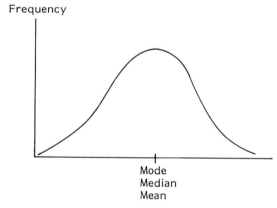

Figure 5.5: Relative positions of the mean, median and mode in a symmetric distribution

Thus, knowledge of the values of any two of the measures of average will indicate something about the skewness of the frequency curve and hence of the underlying frequency distribution.

MEMO

- If the frequency distribution is left- (negatively-) skewed, then the mean is less than the median which in turn is less than the mode.
- If the frequency distribution is right- (positively-) skewed, the mean is greater than the median which in turn is greater than the mode.
- If the frequency distribution is symmetric, then all three measures will be identical.

There are several other measures of average which are used less frequently. Examples are the *trimmed* mean (which attempts to correct the sensitivity of the mean to outliers by omitting say 5 or 10% of the values at each end of the distribution and *then* calculating the mean), the *geometric* mean (used for calculating average rates of growth over time), and a number of measures in a field of statistics known as "exploratory data analysis". Space considerations prevent further discussion of these alternatives.

PERCENTILES

Before we conclude this chapter, there is one final measure in the descriptive statistics toolbox which we must consider. While it is a measure of location rather than of average, it enables us to identify or locate particular values in the sample which define specified percentages of the whole sample.

The median is the value which divides the n sample values, after they have been arranged in ascending order, into two equal parts with 50% of the sample values below the median and 50% above it. It is often very useful to be able to determine the location of values other than the 50% value. For example, we might want to identify the sample value above which 10% of the values lie, or below which 5% lie, and so on. These values which identify any chosen percentage of the whole sample are known generally as *percentiles*.

So for example the value below which 25% of the sample values lie is known as the 25th percentile. The value below which 80% of sample values lie is the 80th percentile. Note that the 25th percentile (below which 25% of the values lie) is the same value *above* which 75% of the values lie and the same reasoning applies to all percentile values. Percentiles are only usually calculated for metric and *numeric* ordinal data.

Recall that to find the median value we first arranged the raw data into ascending order, numbering the position of each value, and then located the middle value (or the mean of the two middle values if n was even). This median value leaves 50% of the sample values to the left (smaller) and 50% to the right (larger). With larger samples the easiest way to find the *position* of this middle observation is to calculate the value of $\frac{1}{2}(n + 1)$, i.e. the value of $(50/100) \times (n + 1)$, and in this sense the median can be interpreted as the *50th percentile*. In exactly the same way, to find the value of, say, the 25th percentile we first determine its position using the expression $(25/100) \times (n + 1)$, and find the value of the sample observation in this position.

As an example, let us find the 25th and 75th percentiles for the 30 ordinal "before" Oswestry Mobility Scale scores shown in Table 5.8 in ascending order. Since $n = 30$ then $(n + 1) = 31$ and:

$$\frac{25}{100} \times 31 = 7.75$$

So the 25th percentile is the value of the observation in the 7.75th position, i.e. three-quarters of the way from the value of the 7th observation to the value of the 8th observation. The 7th observation has a value of 15 and the 8th a value of 25, so the 25th percentile is three-quarters of the way from 15 to 25, i.e. is 22.5 (you'll need to do a spot of simple arithmetic here). In other words, one-quarter of the patients in the sample had mobility scores of 22.5 or less.

For the 75th percentile:

$$\frac{75}{100} \times 31 = 23.25$$

So the 75th percentile is the value of the observation a quarter of the way from the 23rd to the 24th observations. The 23rd observation has a value of 95 and the 24th a value of 100. So the 75th percentile is 96.25. In other words, three-quarters of the patients in the sample had a mobility score of 96.25 or less.

We will find in the next chapter that the 25th and 75th percentiles are of particular interest. They are generally known as the first and third *quartiles*, and usually denoted Q1 and Q3 respectively, because, along with median (which in turn is denoted Q2), they divide the sample observations—and thus the frequency distribution—into four equal parts. Each part contains one quarter, $n/4$, of the observations in the sample. This idea is illustrated in Figure 5.6.

MEMO

- The 25th and 75th percentiles are known as the first and third quartiles.
- The median is the second quartile.
- These three values divide the frequency distribution into four equal parts.

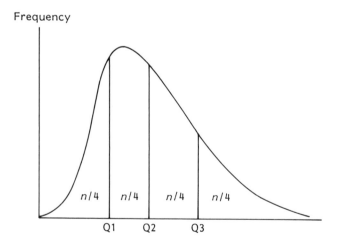

Figure 5.6: The three quartiles divide the area under the total frequency curve (and the total number of observations n) into four equal parts

An Example from Practice

Table 5.12 contains information on premature mortality and long-standing illness[3], expressed in comparison with a "normal" score of 100, among a sample of individuals aged less than 65. So scores higher than 100 indicate higher than normal mortality or higher than normal long-standing illness. The authors were studying the relationship of mortality and long-standing illness to deprivation. Deprivation was measured by the Townsend Deprivation Score*, and these scores were used to divide the sample subjects into deprived fourths or quartiles.

The results indicate that the more deprived the subject, the greater the likelihood of premature mortality or long-standing illness, and "the difference in long-standing illness between the most and least deprived fourths of the sample was larger than the difference in premature mortality". The authors suggest that long-standing illness ratios might be a better basis for the allocation of resources than premature mortality ratios.

Table 5.12:   Illustrating the use of quartiles: mortality and long-standing illness ratios versus levels of deprivation

QUARTILE	MORTALITY RATIO	LONG-STANDING ILLNESS RATIO
Most deprived quarter	108	128
Second most deprived quarter	86	92
Third most deprived quarter	87	85
Least deprived quarter	83	80

Reference will often be seen to *deciles*. These divide the sample observations into 10, rather than 100, equal parts. In other words the bottom, or first, decile has the same value as the 10th percentile, the 7th decile the same as the 70th percentile, etc. The decile values can thus be calculated using the procedure described above, first converting deciles into the corresponding percentiles.

USING A COMPUTER TO FIND MEASURES OF AVERAGE

Even fairly basic hand calculators, as long as they have a statistics facility, will easily calculate a mean (and a "standard deviation" which we will come to in the next chapter). Since each calculator is different, I obviously cannot give instructions as to how to make these calculations—readers will have to consult the instruction booklet that comes with their model. For anything more than these two measures it will usually be easier to use one of the major statistical packages, i.e. SPSS, Minitab, UNISTAT, etc. or a spreadsheet such as Microsoft Excel.

---

* The Townsend Deprivation Scale measures comparative deprivation by scoring subjects on four factors. The Jarman Scale performs a similar role.

Using SPSS to find measures of average

Finding values for the three measures of average and any required percentile value is straightforward. We can use the "before" and "after" Oswestry Mobility Scale scores in Table 3.6. The data are ordinal but neither SPSS nor Minitab distinguishes data types and each program produces all three summary measures regardless of type. If the Oswestry data are entered into columns c1 and c2, then the following commands will produce a full range of summary statistics:

> **Statistics**
> > **Summarize**
> > > **Frequencies**
> > > > **Select c1 and c2**
> > > > > **Statistics**
> > > > > > ⊠ **Quartiles**
> > > > > > > ⊠ **Mean**
> > > > > > > > ⊠ **Median**
> > > > > > > > > ⊠ **Mode**
> > > > > > > > > > ⊠ **Minimum**
> > > > > > > > > > > ⊠ **Maximum**
> > > > > > > > > > > > ⊠ **Skewness**
> > > > > > > > > > > > > ⊠ **Kurtosis**

The resulting output is shown in Figure 5.7.

```
BEFORE

Mean        61.267    Median    66.500    Mode          10.000
Kurtosis   -1.433     SE Kurt    0.833    Skewness    -0.167
SE Skew      0.427    Minimum   10.000    Maximum    115.000


Percentile  Value     Percentile  Value    Percentile  Value
25.00       22.500    50.00       66.500   75.00       96.250


Valid cases  30

AFTER

Mean        72.500    Median    72.500    Mode          40.000
Kurtosis   -0.857     SE Kurt    0.833    Skewness    -0.049
SE Skew      0.427    Minimum   15.000    Maximum    132.000


* Multiple modes exist. The smallest value is shown.


Percentile  Value     Percentile  Value    Percentile  Value
25.00       47.500    50.00       72.500   75.00      100.000


Valid cases  30
```

Figure 5.7: Descriptive summary statistics produced by SPSS for the Oswestry Mobility Scale "before" and "after" scores (Table 3.6)

The output includes an ungrouped frequency, relative frequency and cumulative frequency table which I have suppressed (by unchecking **Frequency Table**) since we have already seen this output in Table 3.7.

Most of the output is self-explanatory. Notice the warning of multiple modes in the "after" output. The program calculates the mean Oswestry score even though the data are ordinal; this should of course be ignored. Standard error is a statistical inferential measure which need not concern us in this book, except to say that the smaller the standard error, the better our sample mean will be as an estimate of the mean of the sampled population.

The kurtosis and skewness coefficients (see Chapter 3) indicate the degree of departure of the distribution of the sample scores from "Normal". Recall that negative skewness means a negatively skewed distribution, and negative kurtosis a distribution more peaky than "Normal". I have not discussed the standard error (SE) of either skew or kurtosis, which should be ignored. The three quartile values are given as the 25th, 50th and 75th percentiles.

Using Minitab to find measures of average

If the Oswestry data are first entered into columns c1 and c2 say, then the following commands will produce a set of descriptive summary measures:

> **Stat**
>> **Basic Statistics**
>>> **Descriptive Statistics**
>>>> **Select c1**
>>>>> **Select c2**
>>>>>> **OK**

The output produced by Minitab is shown in Figure 5.8.

```
MTB > Describe ''Before'' ''After''.

Descriptive Statistics

Variable   N       Mean      Median    TrMean    StDev     SEMean
Before     30      61.27     66.50     61.19     37.20     6.79
After      30      72.50     72.50     72.62     31.23     5.70

Variable   Min     Max       Q1        Q3
Before     10.00   115.00    22.50     96.25
After      15.00   132.00    47.50     100.00
```

Figure 5.8: Descriptive summary statistics produced by Minitab for the Oswestry Mobility Scale "before" and "after" scores (Table 3.6)

Again the output is reasonably self-explanatory. There is no choice of output options in Minitab as there is in SPSS. Notice that the mode is not calculated. The first and third quartiles are presented as Q1 and Q3. The trimmed mean (TrMean) is the mean calculated after the smallest and largest 5% of values have been omitted. However, these data are ordinal so the mean values should be ignored anyway. StDev stands for "standard deviation" and is a measure of spread which I will deal with in the next chapter. It doesn't seem that Minitab can calculate percentiles.

Using Microsoft Excel to find measures of average

If the Oswestry "before" data from Table 3.6 are entered into column A of the Excel spreadsheet, with the first cell used for the label "Before", then the following commands will produce a set of descriptive statistics:

> **Tools**
>> **Analysis Tools**
>>> **Descriptive Statistics**
>>>> **Input Range** (type $A:$A:)
>>>>> **Output Range** (type $B:$B:)
>>>>>> ☒ **Label in First Row**
>>>>>>> **Grouped by:** ⊙ **Columns**
>>>>>>> **OK**

The output is shown in Figure 5.9.

```
Single descriptive statistics (column)

Performed By:   ISD
Date: 28-July-95 at 8:54
Range: [Book1]Sheet1: A [1 - 31]

n:            30            Minimum:              10
Sum:          1838.0        25th Percentile:      25
Mean:         61.267        Median:               66.5
Variance:     1384.2023     75th Percentile:      95
SD:           37.2049       Maximum:              115
SE:           6.7926
Skewness:     -0.1674       Mode:                 10
Kurtosis:     -1.433

95% CI of Mean:
Minimum:   47.3741
Maximum:   75.1592

Range                      Midpoint    Frequency    Cumulative %
9.0 to   26.9999999999     18.0        9            30.0
27.0 to  44.9999999999     36.0        2            36.7
45.0 to  62.9999999999     54.0        3            46.7
63.0 to  80.9999999999     72.0        6            66.7
81.0 to  98.9999999999     90.0        3            76.7
99.0 to 116.9999999999     108.0       7            100.0
```

Figure 5.9: Output from the Excel Descriptive Statistics program for the Oswestry Mobility Scale "before" scores (Table 3.6)

The output is quite comprehensive and reasonably self-explanatory. Excel first gives the sample size, $n = 30$. Sum = 1838.0 is the sum of the scores. The three measures of average are calculated; the mean as 61.267, the median as 66.5, and the mode as 10. Standard deviation and variance are measures of spread which we will deal with in the next chapter. The 95% CI of Mean is a statistical inferential measure which we cannot deal with here, except to say that the wider the confidence interval (CI) the less accurate is our sample mean as an estimate of the mean of the sampled population.

The kurtosis and skewness coefficients (both negative) indicate the degrees of departure of the distribution of the sample scores from "Normality". Recall that negative skewness means a negatively skewed distribution, and negative kurtosis a distribution more peaky than Normal (see Chapter 3). The maximum and minimum sample scores are given as 10 and 115, as are the 25th and 75th percentiles (i.e. the first and third quartiles). The 95% CI of the mean is a statistical inferential measure which we will discuss later. Excel also produces a grouped frequency distribution with the class midpoints and percentage cumulative frequency given.

Using EPI Info to find measures of average

If the Oswestry data are entered into the *before* and *after* fields of a file called "osw.rec", then the following command sequence will produce a set of summary descriptive measures:

> **Programs**
> > **Analysis**
> > > **read osw.rec**
> > > > **freq before after**

Alternatively the command **means** can be used instead of the command **freq**, the output is the same. The problem with EPI, of course, is that the preliminary procedures, creating a "qes" file, then creating the corresponding "rec" file, have to be gone through. If these files have already been set up for other purposes then the use of EPI may be convenient; if not, SPSS, Minitab or Excel will produce results more quickly. As a matter of interest the output from the above commands is shown in Figure 5.10.

EPI does not allow any choice in what is output, and by default it also produces a frequency and cumulative frequency table which I have suppressed. Otherwise the output is very similar to that of Minitab, except that EPI also calculates a value for the mode.

SUMMARY

In this chapter we have looked at a number of ways in which we can summarise sample data in terms of its average or location.

A measure of average (or average or central tendency) is the first of two important measures used by statisticians to summarise frequency distributions. The measure of average provides us with a quantitative idea of the "centre" of the

```
BEFORE

Total    Sum      Mean     Variance   Std Dev    Std Err
30       1838     61.267   1384.202   37.205     6.793

Minimum  25%ile   Median   75%ile     Maximum    Mode
10.000   25.000   66.500   95.000     115.000    10.000

AFTER

Total    Sum      Mean     Variance   Std Dev    Std Err
30       2175     72.500   975.017    31.225     5.701

Minimum  25%ile   Median   75%ile     Maximum    Mode
15.000   50.000   72.500   100.000    132.000    40.000
```

Figure 5.10:   Descriptive summary statistics produced by EPI for the Oswestry Mobility
Scale "before" and "after" scores (Table 3.6)

distribution, the value around which a good portion of the values in the distribution will be clustered.

There are three commonly used measures of average, and the type of variable being analysed determines which measure is most suitable. The mean is a measure of conventional average and is generally used with metric data; it should not be used as a measure of average for ordinal data. The median locates the central value and is generally used with ordinal data (but also with skewed metric data). The mode, which measures the typical value, is the only measure of average possible if the data are nominal.

In this chapter we also discussed percentiles (or centiles) which locate the position of the sample value below or above which some percentage of the sample values will lie.

The second important summary measure describes the degree to which the values are spread out, whether they are narrowly or widely dispersed. Such measures are known as *measures of spread* or *dispersion*. In the next chapter we will discuss several of these measures.

EXERCISES

5.1 (a) Calculate the mean, median and mode of the following scores:

     1 2 2 3 4 4 4 8 8

   (b) Recalculate the three measures after: (i) adding 2 to each score; (ii) subtracting 2 from each score; (iii) multiplying each score by 2; (iv) dividing each score by 2.

   (c) Write down some general rules governing the consequences for each measure of average of adding, subtracting, multiplying and dividing scores by a constant.

5.2 The data in Table 5.13 contain the scores given by two observers to the same 16 patients admitted with trauma of varying degrees of seriousness[4]. The observers were using the Injury Severity Scale which ranges from 0 (no injury) to 100 (death-inducing injury).

(a) What type of variable is involved here?
(b) Arrange each of the two sets of ISS scores in ascending order.
(c) Explain why you might hesitate before using the mean as a measure of average for the ISS score given by each observer.
(d) Calculate and compare the median and modal ISS scores for Observers 1 and 2. Interpret your results.

Table 5.13:   Injury Severity Scores given by two observers to 16 patients admitted with trauma

CASE NUMBER	OBSERVER 1	OBSERVER 2
1	13	9
2	22	13
3	29	27
4	20	16
5	22	22
6	19	17
7	29	54
8	13	13
9	34	29
10	4	4
11	25	75
12	18	35
13	9	4
14	50	50
15	9	14
16	34	50

5.3 Refer back to the GP patient waiting times in Table 4.9.

(a) What type of variable is involved here? What measures of average are appropriate?
(b) Calculate the appropriate measures of average for waiting times. Interpret your results in terms of how each measure captures the idea of average or central tendency.

5.4 In a study of the efficacy of malathion and d-phenothrin as treatments for human lice and nits in children[5], the researchers measured hair length and hair colour of the subjects. These are shown in Table 5.14.

(a) What types of variable are these?
(b) Calculate appropriate measures of average for hair length and colour. Interpret your results.

5.5 Refer back to Table 3.14 which contains Waterlow pressure-sore risk scores for a sample of 40 patients.

Table 5.14: Hair length and colouring of subjects in a study on effectiveness of malathion and *d*-phenothrin on human lice and nits

LENGTH OF HAIR	NUMBER OF CHILDREN	
	Malathion treated	*d-phenothrin treated*
Long	37	20
Mid-long	23	33
Short	35	44

LENGTH OF HAIR	NUMBER OF CHILDREN	
	Malathion treated	*d-phenothrin treated*
Blond	15	18
Brown	49	55
Red	4	4
Dark	27	21

(a) What type of variable is the Waterlow scale?

(b) What measures of average would you consider appropriate for these data?

(c) Calculate the median and modal Waterlow scores. Interpret your results.

(d) Calculate the 10th and 90th percentile Waterlow scores and interpret.

(e) Twenty-five per cent of patients in the sample had Waterlow scores (i) greater than what value? (ii) less than what value?

(f) Draw the ogive for the Waterlow scores and check the results of your calculations in (c) for the median and in (d).

5.6 Refer back to the data in Table 4.9 showing the time spent by 60 patients waiting to see their GP at a large health centre.

(a) Calculate the mean and median waiting times. Comment on your results.

(b) Calculate the second and eighth deciles for waiting times. Comment on your results.

(c) Draw the ogive and confirm your results in (b).

5.7 The data in Table 5.15 record the response of junior doctors to the amount of time spent carrying out inappropriate tasks.

(a) What type of variable is involved here?

(b) What measures of average would you consider most suitable for summarising these data? Calculate it or them. Comment on your result(s).

5.8 The data in Table 4.4 record the duration (in seconds) of first and fourth seizures for 40 patients receiving at least four ECT treatments.

(a) Can you identify any outliers in the raw sample data?

(b) Rearrange each column in the data into ascending order.

(c) Can you discern any broad patterns in the data? What value would you guess the data clusters or congregates around in each column?

Table 5.15: Frequency distribution on the response of junior doctors on amount of time carrying out inappropriate tasks

INAPPROPRIATE TIME	PERCENTAGE OF JUNIOR DOCTORS
Reasonable	20
Acceptable	21
Moderate	23
Excessive	26
Grossly Excessive	10
Total	100

(d) Calculate the mean, median and mode values for both first and fourth seizure durations. Interpret your results.

(e) Comment on the advantages and limitations of each measure of average with these data.

(f) Calculate the 5th and 95th percentile values for first and fourth seizure durations.

(g) Draw the ogives for first and fourth seizure durations. Use these to check your results for the median value and from (c).

(h) Use the ogives to answer the questions (i) "A fifth of patients had seizures lasting longer than how many seconds?"; (ii) "A third of patients had seizures lasting less than how many seconds?".

5.9 The raw data in Table 4.8 record the number of smokers who are co-workers of the subjects of the study into passive smoking and heart disease.

(a) What type of variable is being measured here?

(b) Arrange the data into ascending order, and comment on any broad patterns you can see.

(c) Calculate cumulative frequency and chart it using a step diagram.

(d) Calculate the three measures of average. Interpret your results.

REFERENCES

1. Tunnicliffe, W. S. *et al.* (1994) Effect of domestic concentrations of nitrogen oxide on airway responses to inhaled allergen in asthmatic patients. *The Lancet*, **344**, 1733–5.

2. Fertig, A. *et al.* (1993) Understanding variation in rates of referral among general practitioners: are inappropriate referrals important and would guidelines help? *BMJ*, **307**, 1467–70.

3. Fowle, S. and Stewart-Brown, S. (1994) Deprivation contributes to chronic illness (letter). *BMJ*, **308**, 204.

4. Gillard, J. H. *et al.* (1993) Pre-registration house officers in eight English regions: survey of quality of training. *BMJ*, **307**, 1180–4.

5. Chosidow, O. *et al.* (1994) Controlled study of malathion and *d*-phenothrin lotions for *Pediculus humanus* var *capitas*-infested schoolchildren. *The Lancet*, **344**, 1724–6.

MEASURES OF SPREAD

NEVER MIND THE QUANTITY, FEEL THE WIDTH

When we want to uncover the characteristics of sample data, i.e. when we want to *describe* the data, we have seen that we can:

- Construct a frequency distribution and maybe also a chart. This will give us a *broad* overview of the principal features of the data.
- Calculate a measure of average (the mean, the median or the mode). This will give us a more precise idea of the value around which the sample values tend to locate or congregate.

Chapters 3 and 4 discussed frequency distributions and charts. In Chapter 5 we looked at measures of average. In addition we will usually also want a measure that summarises the degree to which the sample data are *spread* or *dispersed*. For example, look at the two charts in Figure 6.1. The mean and median of the two frequency distributions are both equal to 5 and yet the distributions are clearly quite different. In the upper figure the data only take the values 4, 5 and 6. In the lower figure the values are much more widely spread, ranging from 1 to 9.

An Example from Practice

We can illustrate the idea of dispersion or spread by referring to the raw sample data in Table 6.1, which shows the number of wound dressings applied by two grades of community nurses. The 11 enrolled nurses together applied 52 wound dressings on the day in question and the nine auxiliary nurses 16. The mean number of dressings applied by the enrolled nurses is $52/11 = 4.7$, and the mean number applied by the auxiliary nurses is $16/9 = 1.8$. So on average the enrolled nurses applied more than twice as many dressings as the auxiliary nurses.

Frequency

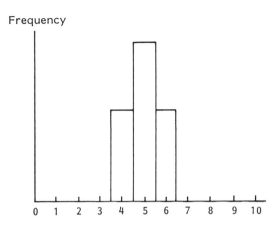

Frequency

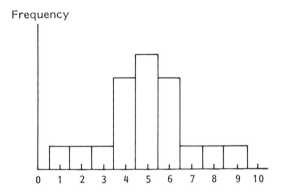

Figure 6.1: Two distributions with same location but different spread

Table 6.1: Number of wound dressings applied in one day by a sample of enrolled and auxiliary nurses

No. ENROLLED NURSE DRESSINGS	A	B	C	D	E	F	G	H	I	J	K
	6	6	4	6	2	2	2	3	7	6	8
No. AUXILIARY NURSE DRESSINGS	L	M	N	O	P	Q	R	S	T		
	1	2	3	1	2	4	1	1	1		

Not only that. The *spread* in the number of dressings applied by each of the enrolled nurses is noticeably wider (varying between 2 and 8), than is the spread in the number of dressings applied by each of the auxiliary nurses. The latter varies only between 1 and 4 (and with *five* of these nurses all applying the same number of dressings, one). One consequence of the narrower spread in the number of dressings applied by the auxiliary nurses is that, on the whole, the number of dressings applied by each of them is quite close to their mean of 1.8. The auxiliary nurse who applied four dressings is furthest away from this mean (a difference of $4 - 1.8 = 2.2$). On the other hand, the *wider* spread in the number of dressings applied by the enrolled nurses means that, on the whole, the number of dressings applied by each of them is not as close to their mean of 4.7. The enrolled nurse who applied eight dressings is furthest away (a difference of $8 - 4.7 = 3.3$). The consequence is that in general the mean for distributions with narrower spreads is more *typical* of (i.e. closer to) the individual sample values than is the mean for data that are more widely spread.

MEMO

Measures of spread or dispersion indicate, roughly speaking, the average distance of the sample values from the centre of the distribution values.

Spoilt for Choice—Again

In this chapter we shall be looking at three different types of measures of spread:

- *Those that are frequency-based*. These measures are based on the way in which the sample values are spread out among the different categories or classes. They are used principally with nominal and non-numeric ordinal variables, and include the variation ratio (VR), the index of diversity (ID), and the index of qualitative variation (IQV).
- *Those that are range-based*. These measures are based on the difference (or range) between the largest values in the sample and smallest. They are used mainly with numeric ordinal variables and skewed metric variables, and include the range, the interquartile range (IQR)and the semi-interquartile range (SIQR).
- *Those that are deviation-based*. These measures are based on the average difference (or deviation) between each sample value and the mean of the sample values, and are used with metric variables. Deviation-based measures include the standard deviation (*s.d.*).

MEMO

There are three types of spread measures: frequency-based; range-based; and deviation-based.

Choosing a Measure of Spread

As with measures of average, the choice of an appropriate measure of spread is dictated by the type of data. The choices are as follows:

- With *nominal* data we would normally be restricted to one of the frequency-based measures (the use of a range-based measure, although possible, would not in general be very helpful).

- With *ordinal* data with alphabetic (i.e. non-numeric) categories we are similarly restricted to one of the frequency-based measures, although a range-based measure might on occasion be used.
- With "numeric"* ordinal data we would normally choose from the range-based measures, although if the number of possible values is small enough not to require a grouped frequency distribution we might use a frequency-based measure.
- With metric variables we would usually employ a deviation-based measure, although, as we will see shortly, with a skewed distribution a range-based measure might be more appropriate.

"That's what you call a measure of spread!"

The shape of the frequency distribution, and whether there are outliers (extreme values) in the sample data, will also need to be taken into account. Table 6.2 summarises these possibilities.

Table 6.2: Measures of spread and types of data

TYPES OF MEASURE	TYPES OF DATA			
	Nominal	*Ordinal non-numeric*	*Ordinal "numeric"*	*Metric*
Frequency-based (VR, ID, IQV)	yes	yes	no	no
Range-based (Range, IQR, SIQR)	no	no	yes	yes
Deviation-based (s.d.)	no	no	no	yes

---

* I have put inverted commas around the word numeric as a reminder that, although they might take a numeric form, we are not talking *real* numbers here.

```
┌─────────────────────────────────────────────────────────────────────┐
│                              MEMO                                      │
│                                                                       │
│   ■ Frequency-based measures of spread are used for nominal and non-   │
│     numeric ordinal data.                                             │
│   ■ Range-based measures of spread are used with numeric ordinal data  │
│     and sometimes with skewed metric data.                           │
│   ■ Deviation-based measures of spread are used with metric data.     │
└─────────────────────────────────────────────────────────────────────┘
```

MEASURES OF SPREAD FOR NOMINAL VARIABLES

Measures of spread are not used as commonly with nominal data as they are with ordinal and metric data[*]. This is unfortunate because, despite their limitations (to be discussed below), they are useful summarising statistics when nominal data are involved, especially in the absence of any alternative.

As Table 6.2 indicates, with nominal variables we are restricted to using one of the frequency-based measures of spread. Frequency-based measures of spread measure the degree of *heterogeneity* of the sample values; in other words, the degree to which the sample values are divided among the categories. At one extreme, if all the sample values fall into just one category, we refer to the spread as *homogeneic*, i.e. the spread is *zero*. At the other extreme, if there are an equal number of values in each category, we would describe the spread as *heterogeneic*, i.e. the sample values are spread out as far as possible.

For example, the frequency distribution in Table 6.3(a), which records the blood type of 12 patients, is a case of homogeneity; since all the sample values fall into a single category, the data are not spread at all.

Table 6.3:   Examples of homogeneic and heterogeneic
frequency distributions

(a) HOMOGENEIC FREQUENCY DISTRIBUTION

Blood type	Number of patients
A	0
B	0
A/B	0
O	12

(b) HETEROGENEIC FREQUENCY DISTRIBUTION

Blood type	Number of patients
A	3
B	3
A/B	3
O	3

---

[*] To the author's knowledge none of the major statistical packages includes calculation of any of the three frequency-based measures of spread discussed here.

In contrast, the frequency distribution in Table 6.3(b) describes a sample that is heterogeneic, because the sample values are spread evenly among all the categories (statisticians call such a distribution *uniform*).

There are three frequency-based measures of spread. We start by looking at the most simple to calculate and understand.

The Variation Ratio (VR)

The variation ratio (or *VR*) measures the proportion of sample observations that do *not* fall in the modal class. The value of the VR when there is no spread (i.e. when all cases fall into just one category) is 0. The *theoretical* maximum value is 1, but this can only be achieved if there is an infinite number of categories with one value in each! In practice the *actual* maximum value (which will always be *less* than 1) depends on both the sample size and the number of categories, but gets closest to 1 when the number of sample values in each category is the same. This dependence of the maximum value of the VR on the number of categories makes it difficult to compare distributions with different numbers of categories, but the VR is a perfectly good measure of spread for similar distributions. We calculate the VR as follows:

- *Step 1:* Construct a frequency distribution for the sample values.
- *Step 2:* Find the mode, i.e. the category with the highest number of values, and hence the frequency of this modal category (see Chapter 5 if you need a reminder).
- *Step 3:* Divide this modal frequency by the number of values in the sample (usually denoted n).
- *Step 4:* Subtract the value obtained in Step 3 from 1. The result is the VR.

The process of calculating the VR can be described algebraically (skip this if you want):

$$VR = 1 - (f_{mode}/n)$$

where f_{mode} is the modal value and n is the sample size.

For example, the VR for the blood types in Table 6.3(a) is found thus:

- *Step 1:* The frequency distribution is already constructed, Table 6.3(a).
- *Step 2:* The modal category is type O, and the modal frequency is 12.
- *Step 3:* The sample size equals 12, so dividing the modal frequency by the sample size gives $12/12 = 1$.
- *Step 4:* The VR is therefore $1 - 1 = 0$. So the proportion of sample values falling *outside* the modal category is zero, and this frequency distribution therefore has zero spread.

For the frequency distribution in Table 6.3(b), following the same procedure, the modal frequency is 3 (there are in fact four modes, all equal to 3), and the VR is thus:

$$VR = 1 - 3/12 = 1 - 0.25 = 0.75$$

This means that 75% of values fall outside the modal category. So although we know that this spread is as wide and even as possible, we don't get a value of 1.

As we noted above, this makes interpretation of any single VR difficult. Clearly the closer to 1 the VR is, the greater the spread; but the VR is best used to *compare* the spread of two or more similar frequency distributions, the higher the value of the VR the greater the spread of sample values among categories.

An Example from Practice

Table 6.4 contains data on a number of nursing activities as a cause of back pain in two hospitals, one in Athens (Greece) and the other in California[1].

Table 6.4: Nursing activities as a cause of back pain in two hospitals

ACTIVITY	NUMBER OF NURSES REPORTING BACK PAIN EPISODES	
	Athens	*California*
Move equipment ≥ 15 kg	146	55
Lift patient onto trolley	130	121
Lifting a patient in bed	118	264
Help patient get out of bed	106	165
Bending to lift object from floor	98	0
Move bed	85	148
None of the above	73	138
Totals	756	891

In Athens, the modal category is "Moving equipment weighing ≥ 15kg", with a modal frequency of 146 out of $n = 756$. This gives a VR of:

$$VR = 1 - 146/756 = 1 - 0.193 = 0.807$$

So 80.7% of the sample values don't fall in the modal category. In California, the modal category is "Lifting a patient in bed" with a modal frequency of 264 out of $n = 891$. Thus:

$$VR = 1 - 264/891 = 1 - 0.296 = 0.704$$

So 70.4% of the sample values don't fall in the modal class. Since 0.807 is closer to 1 than 0.704, then the spread of values across the categories is greater in the Athens data than in the California data.

It is not possible to place any precise interpretation on any numeric value we obtain for the VR of any given sample. All we can say is that low values of VR (i.e. values towards zero) indicate lower amounts of spread (most sample observations in just a few categories, some categories perhaps empty), while high values of VR (i.e. values towards 1) indicate higher amounts of spread (similar numbers of values in every category). It is not easy to say what a value of, say, 0.4 or 0.7 for an individual sample means. For this reason, as noted above, the VR is often more useful for comparing the spread of two or more sample distributions.

In terms of desirable properties, the VR does not use all of the information in the sample (based as it is on the mode), and is sensitive to small changes in the data. However, it is easy to understand and calculate, and is resistant to outliers.

MEMO

The variation ratio measures the proportion of sample values that do not fall in the modal category. The closer the VR is to 1, the greater the spread of the sample values. The maximum value of the VR depends on the number of categories.

The Index of Diversity (ID)

Whereas the VR considers only the frequency of the modal class and ignores the remaining sample values, the index of diversity (or *ID*) is based on the proportion of sample observations in *each* category. At one extreme, the ID is zero if all values fall in the same category (zero spread); at the other extreme, the ID *approaches* its theoretical maximum value of 1 when each observation is in a separate category (i.e. maximum spread). In any particular case its actual maximum value is equal to the number of categories minus 1, divided by the number of categories. So like the VR, the maximum value of the ID depends on the number of categories in the distribution. This makes it unsuitable for comparing the spread in distributions with differing numbers of categories.

We calculate the ID as follows:

- *Step 1:* Construct a frequency distribution to determine the frequency of each category.
- *Step 2:* Divide the frequency of the first category by n, the total number of values in the sample, to give the proportion of cases in the first category. Repeat for each category.
- *Step 3:* Square each of the values found in Step 2 and add all the resulting values together.
- *Step 4:* Subtract the final sum in Step 3 from 1. The result is the ID.

We can express the procedure for calculating the index of diversity algebraically as follows (skip this if you wish):

$$\text{ID} = 1 - (p_1^2 + p_2^2 + \ldots + p_k^2)$$

where p is the proportion of observations in each category and k is the number of categories. Hence, to return to the nurses' back pain example, the ID for Athens is:

- *Step 1:* See Table 6.4 for the frequency distribution.
- *Step 2:* Dividing each category frequency by total frequency of 756 gives:

  146/756   130/756   118/756   106/756   98/756   85/756   73/756

  Which gives the proportions:

  0.1931   0.1720   0.1561   0.1402   0.1296   0.1124   0.0966

- *Step 3:* Squaring each of these values and adding the squared values together gives:

  $0.1931^2 + 0.1720^2 + 0.1561^2 + 0.1402^2 + 0.1296^2 + 0.1124^2 + 0.0966^2 = 0.1496$

■ *Step 4:* Subtracting this value from 1 gives the index of diversity:

$$ID = 1 - 0.1496 = 0.8503$$

With seven categories the maximum ID value is equal to $(7 - 1)/7 = 6/7 = 0.8571$, so the value of 0.8503 implies a spread of values close to the maximum possible. We face the same problem of precise interpretation of any specific numeric value as we did with the VR above, and it is perhaps more interesting to compare this result with the ID = 0.8040 for the California nurses. Since the Athens value of 0.8503 is closer to 1, this indicates that the Athens values are more spread out among the categories than the California values. This confirms the VR result above.

Because it is based on the squares of the proportions, the index of diversity gives more emphasis to the larger frequency values and thus in essence measures the degree of *concentration* of the sample values into a few large categories.

MEMO

The index of diversity measures the degree of concentration of sample values into a few large categories. The maximum value of the ID depends on the number of categories.

The Index of Qualitative Variation (IQV)

The *IQV* is based on the same idea as the index of diversity except that it overcomes the limitation of the latter measure's uncertain maximum value. The IQV has a value of 0 when all the sample values fall in the same category, and a maximum value of 1, when the number of values is the same in every category. Knowing that the maximum value is always 1 regardless of the number of categories enables us to make a better assessment of any given value we obtain for the index. It also means we can compare spreads between frequency distributions with different numbers of categories. We calculate the value of the IQV as follows:

■ *Step 1:* Calculate the Index of Diversity following the steps outlined above.
■ *Step 2:* Divide the result for the ID obtained in Step 1 by the number of categories minus 1.
■ *Step 3:* Multiply the result obtained in Step 2 by the number of categories. The resulting value is the IQV.

We can express the above procedure for calculating the IQV algebraically as follows (skip this bit if you wish):

$$IQV = [1 - (p_1^2 + p_2^2 + \ldots + p_k^2)]/(k-1)k$$

For the nurses' back pain example, the Athens nurses had an ID = 0.856. Therefore:

■ *Step 1:* The ID = 0.8503

- *Step 2:* The number of categories is seven, so we need to divide the ID by six, which gives:

    0.8503/6 = 0.1417

- *Step 3:* Multiplying the value obtained in Step 2 by the number of categories gives the index of qualitative variation:

$$IQV = 0.1417 \times 7 = 0.9992$$

This value is very close to the known maximum value of 1, indicating a virtually maximum spread of the sample values across categories. Using the same procedure, the IQV for the California nurses is 0.9831, confirming the results from both the VR and the ID above, that the spread of values across categories with the Athens nurses is greater than the spread for the Californian nurses (but not by much).

In terms of desirable properties, the index of diversity and the index of qualitative variation are fairly easy to understand, and not too difficult to calculate. They are also resistant to outliers and use all of the information in the sample.

MEMO

The index of qualitative variation measures the degree of concentration of the sample values into a few categories, as does the index of diversity. However, whereas the ID has a maximum value dependent on the number of categories, the IQV always has a maximum value of 1.

MEASURES OF SPREAD FOR ORDINAL VARIABLES

The three frequency-based measures of spread, VR, ID and IQV, are also the most suitable measures of spread for *non-numeric* ordinal variables.

An Example from Practice

Table 6.5 records the adequacy of information, on both the treatment and about the operation, given to breast cancer patients in hospital, as judged by nurses (in per cent)[2]. A glance at the table indicates that the spread across categories is broadly similar for information given on treatment as that for information given on the operation.

Table 6.5: The adequacy of information given to 125 breast cancer patients in hospital, as judged by nurses

	NONE	LITTLE	MUCH	VERY MUCH
Information on treatment (%)	14.4	44.4	36.7	4.4
Information on operations (%)	36.8	50.5	9.5	3.1

Using the procedures described above we can calculate the VR, ID and IQV values to be:

Information on the treatment:

$$VR = 1 - 44.4/100 = 0.556$$
$$ID = 1 - (0.144^2 + 0.444^2 + 0.367^2 + 0.044^2) = 1 - 0.354 = 0.645$$
$$IQV = 0.645/0.75 = 0.860$$

Information about the operation:

$$VR = 1 - 50.0/100 = 0.495$$
$$ID = 1 - (0.368^2 + 0.505^2 + 0.095^2 + 0.031^2) = 1 - 0.400 = 0.600$$
$$IQV = 0.600/0.75 = 0.800$$

The number of categories equals 4, so the maximum possible value for the ID is $(4 - 1)/4 = 0.75$. Hence on all three measures the spread of values for information given on treatment is slightly greater than is the spread of values on information given about the operation.

For *numeric* ordinal data, we can use range-based measures of spread, in particular the *range* and *interquartile range (IQR)*.

The Range

The *range* is the simplest measure of spread for ordinal data and is easily calculated, being the difference between the smallest and largest values in the sample. So, to determine the range the smallest and largest values in the sample data are first identified and the former subtracted from the latter.

An Example from Practice

Table 6.6 records[3] the Injury Severity Scale (ISS) scores given to 16 patients by two observers. The smallest and largest values (shown bold) are 4 and 45 for Observer 1,

Table 6.6:   Injury Severity Scale scores given to 16 patients by two observers

PATIENT NUMBER	OBSERVER 1	OBSERVER 2
1	9	9
2	14	13
3	29	**75**
4	17	25
5	34	27
6	17	17
7	38	33
8	13	9
9	29	20
10	**4**	**4**
11	29	29
12	25	41
13	4	9
14	16	25
15	25	13
16	**45**	41

and 4 and 75 for Observer 2. This means that the ranges are:

For Observer 1:   range = 45 − 4 = 39
For Observer 2:   range = 75 − 4 = 71

This indicates that the spread of ISS scores given by Observer 2 to the same set of patients is considerably wider than the spread in the ISS scores given by Observer 1.

Although the range is very easy to calculate, it is very sensitive to extreme values or outliers. For example, if the values for patient number 3 are omitted (for which there is an outlier score of 75 given by Observer 2), the ranges now become:

Observer 1:   range = 41 − 4 = 37
Observer 2:   range = 45 − 4 = 41

It would now seem that the spread of scores is narrower for Observer 1 than for Observer 2!

MEMO

The range is the difference between the largest and smallest sample values.

The Interquartile Range (IQR)

One way around the problem of the range's sensitivity to outliers is to chop off some given number—or more usually, some given percentage—of the observations at both the lower and upper ends of the sample distribution. This eliminates the influence of any extreme values, and we can then use the difference between the smallest and largest of the remaining values as the measure of spread.

If we chop off the lowest and highest 25% of the values and measure the difference between the smallest and largest of the remaining values, we get a measure of spread known as the *interquartile range* or *IQR* (sometimes also known as the *midhinge* value). We noted in Chapter 5 that the sample values which identify the lowest and highest 25% of values are known as the first and third quartiles, usually denoted Q1 and Q3 respectively, so the interquartile range is the difference between these two quartile values. In other words:

IQR = value of 3rd quartile − value of 1st quartile

In other words, the IQR provides a measure of the spread of the middle 50% of the sample values. The idea is illustrated in Figure 6.2.

For example, for the Oswestry "before" mobility scores of Table 5.7, we have previously calculated Q1 = 22.5 and Q3 = 96.2. Therefore:

IQR = Q3 − Q1 = 96.2 − 22.5 = 73.7

For the "after" mobility scores, Q1 = 47.5 and Q3 = 100.0, thus:

IQR = 100.0 − 47.5 = 52.5

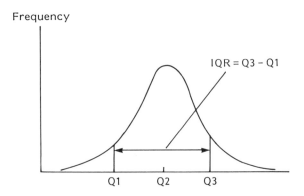

Figure 6.2: Illustrating the idea of the interquartile range

This shows that the spread of Oswestry scores is wider before treatment than after (73.7 compared with 52.5); i.e. in terms of their mobility, the "before" group mobility scores are more spread out or heterogeneic than the "after" scores.

The interquartile range is resistant to outliers; it is also fairly easy to calculate and understand. However, it might be criticised on the grounds that there is nothing magical about the 25% and 75% values—why not, for example, use the lowest and highest 10% or 20% values? Nevertheless, the interquartile range is a popular and oft-quoted statistic.

MEMO

The interquartile range measures the difference between the lowest 25% of sample values (the first quartile) and the highest 25% of sample values (the third quartile).

The semi-interquartile range

The *semi-interquartile range* (known also as the *quartile deviation* or *quartile range*) equals the IQR divided by 2; i.e.

$$SIQR = IQR/2$$

The reason for the division by 2 is to provide a measure of how far the quartiles typically differ from the median. For example, the median (Q2) Oswestry mobility score for the "before" group was calculated in Chapter 5 to be 85. Thus the difference between the median Q2 and the first quartile is:

$$Q2 - Q1 = 85 - 22.5 = 62.5$$

and the difference between the median and the third quartile is:

$$Q3 - Q2 = 96.2 - 85 = 11.2$$

This emphasises the fact that the median is not necessarily half-way between the first and third quartiles (except if the distribution is symmetric). The average difference between the median and the two quartiles is the mean of these two distances, i.e. $(11.2 + 62.5)/2 = 36.8$. The semi-interquartile range shares the properties of the interquartile range but has more of a feeling of a typical deviation from the median. It is becoming as popular as the interquartile range.

MEMO

The semi-interquartile range is the average distance of the first and third quartiles from the median.

MEASURES OF SPREAD FOR METRIC VARIABLES

With metric variables, *deviation*-based measures of spread are generally used, unless the data are markedly skewed or have outliers, in which case the median and interquartile range may produce a more representative set of values. The differences or deviations between each sample value and the sample mean are known as the *mean deviation* values.

The Mean Deviation

The rationale for the use of the deviations as a basis for a measure of spread is that the wider the spread of sample values, on average the further away the values will be from the mean. That is, on average the mean deviations will be larger and the *sum* of these mean deviations will also be larger and can thus be used as a basis for measuring spread. The larger the sum of the mean deviations, the larger is the spread in the sample values.

To illustrate this idea, let us go back to the example at the beginning of this chapter on the number of wound dressings applied in one day by enrolled and auxiliary nurses (see Table 6.1). We calculated the mean number of dressings applied to be 4.7 and 1.8 respectively. If we subtract these mean values from each individual value we get the mean deviation values shown in Table 6.7.

Since the spread of values for enrolled nurses is wider than that for auxiliary nurses, we expect that the mean deviations will also be larger, and this is indeed what we find. For the auxiliary nurses only two of the mean deviations are greater than 1, whereas for the enrolled nurses all except one value are greater than 1. We might expect then that we could use the average of these deviations as a measure of spread; but unfortunately, the average of any set of deviations around the mean is always zero. The reason for this is that the negative and positive values cancel each other out and the *sum* is always zero, so this idea doesn't work.

One way around this would perhaps be to ignore the signs of the mean deviation values, and add up just the values without the minus signs. Such values are known as *absolute* deviations. If we calculate the sum of these absolute deviations for the two grades of nurse we get values of 21.3 and 7.8 respectively for the enrolled and auxiliary nurses. These values confirm the wider spread of the

Table 6.7: Mean deviation values* for number of wound dressings applied by enrolled and auxiliary nurses

ENROLLED NURSE	NO. OF DRESSINGS	MEAN DEVIATION	AUXILIARY NURSE	NO. OF DRESSINGS	MEAN DEVIATION
A	6	1.3	L	1	–0.8
B	6	1.3	M	2	0.2
C	4	–0.7	N	3	1.2
D	6	1.3	O	1	–0.8
E	2	–2.7	P	2	0.2
F	2	–2.7	Q	4	2.2
G	2	–2.7	R	1	–0.8
H	3	–1.7	S	1	–0.8
I	7	2.3	T	1	–0.8
J	6	1.3			
K	8	3.3			

* Mean deviation = number of dressings minus the mean

enrolled nurses' number of dressings. However, for various reasons (including a lack of any sound mathematical properties) the sum of the absolute deviations is not much used as a measure of spread.

There is one other way of overcoming the cancelling effect of the negative and positive deviation values *which leads to the most important measure of spread used with metric variables.*

Standard Deviation

As we have just seen, the idea behind using the average of the mean deviations as a measure of spread doesn't work because of the cancelling of the negative and positive values, so that when we try to add them up before dividing by n (the number of sample values) we always get a zero sum. Taking the average of the absolute deviations is one way around this problem but leads to a bit of a mathematical dead-end. Another way to overcome the cancelling effect is to *square* the mean deviations first, which gets rid of the negative values, then take the average by dividing the sum of the squared deviations by $(n - 1)^*$, and finally take the square-root to get back to the original units.

The end result of this process is a measure of spread known as the *sample standard deviation*, often abbreviated to s.d. or SD and denoted by the symbol s. When there is no spread in the data the standard deviation equals 0, and its value increases as the spread in the data increases, with no upper limit. The standard deviation is rarely calculated by hand, since most calculators have a facility for doing this†, but the necessary steps are as follows:

---

* Why we divide by $n - 1$ to find the average of the squared deviations and not n is connected to the use of the sample standard deviation as an estimate of population standard deviation and need not concern us here. Suffice to say that if we use n we get a resulting value that is too small.
† When calculating the s.d. with a calculator it is important to choose the σ_{n-1} or s_{n-1} key.

- *Step 1:* Calculate the value of the sample mean.
- *Step 2:* Subtract this mean value from each observation (to produce the *mean deviation* values).
- *Step 3:* Square each of these mean deviation values.
- *Step 4:* Add these squared values together.
- *Step 5:* Divide this sum by $(n-1)$.
- *Step 6:* Take the square-root of the value obtained in Step 5. The result is the standard deviation.

Table 6.8 shows this procedure applied to the number of wound dressings applied by the enrolled nurses. The third column contains the mean deviation values (Step 2). The fourth column shows the square of the mean deviations (Step 3). The fourth column is summed to give a value of 48.19 (Step 4), and this is divided by $n-1$, i.e. by $11-1$ (Step 5). This gives a value of 4.819 which is square-rooted (Step 6) to produce a value for the standard deviation of 2.20 dressings. A similar calculation for the auxiliary nurses produces a standard deviation of 1.09 dressings.

Table 6.8: Calculation of standard deviation for number of dressings applied by enrolled nurses

ENROLLED NURSE	NO. OF DRESSINGS	MEAN DEVIATION	MEAN DEVIATION SQUARED
A	6	1.3	1.69
B	6	1.3	1.69
C	4	−0.7	0.49
D	6	1.3	1.69
E	2	−2.7	7.29
F	2	−2.7	7.29
G	2	−2.7	7.29
H	3	−1.7	2.89
I	7	2.3	5.29
J	6	1.3	1.69
K	8	3.3	10.89
Total			48.19

So the standard deviation for enrolled nurses is a little over twice that for auxiliary nurses, indicating that the spread in the number of dressings applied by enrolled nurses is twice as wide as that for auxiliary nurses. Notice that the standard deviation is measured in the same units as the original variable (i.e. number of dressings in this example). This is always the case. But can we say what exactly the values of 2.20 and 1.09 mean?

Interpretation of Standard Deviation

Unfortunately the standard deviation is not as easy to interpret as, say, the variation ratio or the interquartile range. In fact, it is quite difficult to give any common-sense meaning to it. At its simplest, the standard deviation shows how closely the sample values are clustered around the mean[*], but it is *not* the average distance of each sample value from the mean, although it is, more or less, the square-root of the average of these squared distances. This is perhaps not very helpful!

"Seems like a fairly standard deviation to me."

All I can say is that the more you calculate and try to interpret standard deviation values, the less you will worry about trying to attach some everyday meaning to the measure, and the more we will accept it for what it is—and that is a measure which quantifies the spread of a metric variable.

An Example from Practice

Using the procedure outlined above we can calculate the standard deviations for the first and fourth seizure durations for the ECT data in Table 4.4, for which we have already calculated the means as 36.6 seconds and 34.1 seconds respectively. The standard deviation values are:

$$s_{first} = 16.6 \text{ seconds}$$
$$s_{fourth} = 14.1 \text{ seconds}$$

Thus the spread of seizure duration values is wider for the first seizure than for the fourth.

Algebraic Formulation of Standard Deviation

You can omit this section first time round if you wish, with no loss of continuity. We can use the summation notation to express standard deviation algebraically. If $\bar{x}$ is the sample mean, and X is any sample value, then $(X - \bar{x})$ is the corresponding mean deviation and $(X - \bar{x})^2$ is the square of that mean deviation. The sum of squared deviations is then:

$$\Sigma(X - \bar{x})^2$$

---

[*] To be honest I think of the standard deviation myself as the average distance of each sample value from the mean, even though this is not true. But it helps!

Dividing by $(n-1)$ and square-rooting gives us an expression for sample standard deviation:

$$s = \sqrt{\frac{1}{n-1} \Sigma(X - \bar{x})^2}$$

The *variance* is equal to the square of the standard deviation, and is a concept more often used in advanced statistics.

Interpreting Standard Deviation when the Frequency Distribution is Normal

We first encountered the idea of the Normal distribution towards the end of Chapter 4 when we noted that it is a smooth, symmetric, unimodal, and bell-shaped curve. Because of its symmetry it has identical values for the mean, median and mode. The Normal distribution is described by a complicated equation (which we don't have to worry about here), but all we need to draw any Normal curve is the value of its mean and standard deviation. It follows, of course, that there are an infinite number of Normal curves, because there are an infinite number of means and standard deviation values! The idea is illustrated in Figure 6.3, which shows three different (fictitious) Normal curves: (a) for the pulse rates of a healthy group of adults; (b) for the heights of a group of male and female nurses; and (c) for the diastolic blood pressures of a group of traffic wardens.

The Normal distribution is important because many frequency distributions in the human and health sciences are Normally distributed (or are approximately so), and by using what are called the *area properties* of the Normal distribution we can answer all sorts of questions about the way the values in the distribution are dispersed. In particular we can use these properties to help in the interpretation of values we may obtain for the standard deviation. These area properties, which apply to *any* Normal distribution, are shown in Figure 6.4. They are:

- About 68% (i.e. about two-thirds) of the area under the Normal curve is within one s.d. either side of the mean. This is the same as saying that 68% of all the *values* in the distribution will lie within one s.d. of the mean.
- About 95% of the area under the Normal curve is within two s.d.s either side of the mean, i.e. 95% of all *values* will lie within two s.d.s of the mean.
- About 99% of the area under the Normal curve is within three s.d.s either side of the mean; i.e. 99% of all *values* will lie within three s.d.s of the mean.

In other words, if the distribution is Normal, virtually all of the values in the distribution (about 99% of them) will lie within plus or minus three s.d.s of the mean. I have used the word "about" in the above paragraphs; we will see shortly how these values originate and what their exact values are.

MEMO

In any Normal distribution about 68% of the values will lie within plus or minus one s.d. of the mean, about 95% of values will lie within plus or minus two s.d.s of the mean, and about 99% of values will lie within plus or minus three s.d.s of the mean.

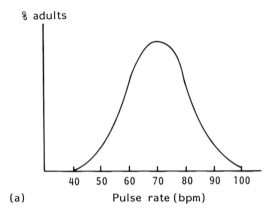

(a)

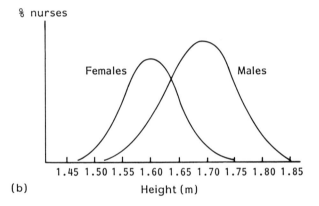

(b)

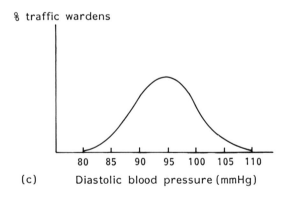

(c)

Figure 6.3:   Three Normal curves

An Example from Practice

Researchers attempting to establish some standard biochemical and physiological responses for healthy men undertaking physical activity took various measurements on a

Frequency

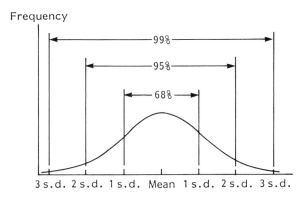

Figure 6.4:   Area properties of the Normal distribution

sample of 20 men while they performed a number of simulated tasks[4]. These measurements included heart beat rate X (in beats per minute, or bpm). The sample mean, $\bar{x}$, and
the sample standard deviation, s, heart beat rate values for the men while they were
performing a wheel-turning exercise were $\bar{x}$ = 123 bpm and s = 18 bpm respectively
(actually 17.9 bpm).

If we assume for the moment that the heart beat rate values were normally distributed, then
the above area properties of the Normal distribution indicate that:

- 68% (approximately two-thirds) of the men had a heart beat rate between the mean
  minus one s.d. and the mean plus one s.d.; i.e. between 123 – 18 and 123 + 18,
  that is between 105 and 141 bpm.
- 95% of the men had heart beat rates between the mean minus two s.d.s and the mean
  plus two s.d.s; i.e. between 123 – (2 × 18) and 123 + (2 × 18), or between 123 –
  36 and 123 + 36, or between 87 and 159 bpm.
- 99% of the men had a pulse rate between the mean minus three s.d.s and the mean
  plus three s.d.s, i.e. between 123 – (3 × 18) and 123 + (3 × 18), or between 69 and
  177 bpm. These values are illustrated in Figure 6.5.

Frequency

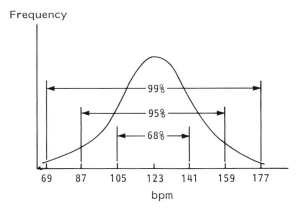

Figure 6.5:   Area properties of the Normal distribution

We can deduce something else about the distribution of values from this study, since the authors also provide minimum and maximum heart rate values for the subjects of 87 and 161 bpm respectively. These values are inside our 99% area property values of 69 to 177, which suggests that the distribution of sample values is not exactly normal. In fact it is probably a little more peaky than the theoretical normal shape (perhaps a bit like the distribution in Figure 6.6).

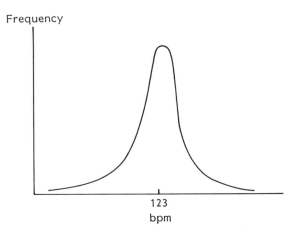

Figure 6.6: A "peaky" distribution

However, it would seem reasonable to assume, because of the nature of the data, that the distribution of heart rate values for the population from which this sample was taken (*all* healthy young men with the same socio-economic characteristics as those chosen for the sample) is normal. In fact the author also provides a value for the median of 121 bpm, which is very close to the mean value of 123 bpm, thus confirming the impression that this distribution is at least symmetric.

Consequently, provided we know the actual minimum and maximum values, we can compare these with the 99% theoretical area property values and this will tell us something about the "normal-ness" or otherwise of the distribution.

These three area values (68%, 95% and 99%) are very useful and worth remembering, but it would be even more useful if we could find the appropriate values for *any* percentages we chose. Indeed we can do this using what is known as the *standard normal distribution*. The discussion over the next few pages is a bit complicated and can easily be omitted if you feel one of your funny turns coming on. Go instead to page 140.

The Standard Normal Distribution

Imagine your friend Vlad, who is on holiday in Norway, telephones you one day and says, "It's quite cold here today, the temperature is zero. What's it like there in the UK?". You respond, "Well, it's 32 here". Vlad exclaims, "Pphew, what a scorcher!". Of course, your friend is using degrees Centigrade, while you are using degrees Fahrenheit. If either one of you were to convert your temperature into the other's scale you would find the temperatures to be the same. As you know, to convert degrees Fahrenheit to degrees Centigrade you would use the *transformation*:

Degrees Fahrenheit = (degrees Centigrade − 32) × 5/9

We can transform the values of any Normal distribution into what is known as the *standard Normal distribution* (we'll see how in a moment) and the great advantage of doing this is that statisticians have already compiled a set of tables showing the proportion of the area under the standard Normal curve for *any* value in the distribution. The standard Normal distribution has the following properties:

- It has the same symmetric bell-shaped curve as any other Normal distribution.
- It has a mean of 0.
- It has a standard deviation of 1.
- The total area under its curve is 1.

A standard Normal distribution is shown in Figure 6.7. Notice that three s.d.s either side of the mean encompasses virtually all of the distribution as with any other Normal distribution.

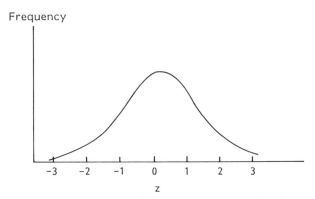

Figure 6.7: Area properties of the *standard* Normal distribution

To see how the transformation works, let us go back to the heart rate example above, assume that heart beat rate is Normally distributed (not unreasonable) and determine the proportion of men in the sample who have a heart rate of between 123 bpm (the mean) and 132 bpm. This is the same as the proportion that the shaded area in Figure 6.8(a) is of the total area under the curve.

We need to transform the values of 123 bpm and 132 bpm into their corresponding standard Normal values. To do this we subtract the sample mean (123) from the value in question and then divide by the standard deviation (18). For the mean value of 123 the transformation is thus:

$$\frac{\text{Value} - \text{mean}}{\text{Standard deviation}} = \frac{123 - 123}{18} = \frac{0}{18} = 0$$

So, as we might have expected, when we transform the mean value of the *original* distribution we get the mean of the standard Normal distribution. The transformation of the value 132 bpm is similarly:

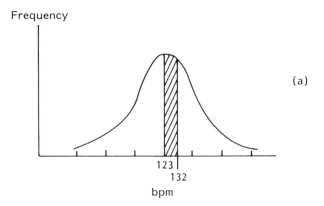

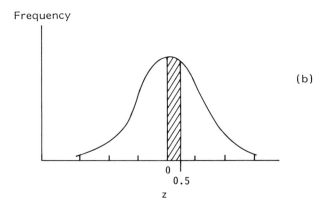

Figure 6.8: Calculating the proportion of men with a heart rate between 123 bpm and 132 bpm, using the standard Normal distribution

$$\frac{132 - 123}{18} = \frac{9}{18} = 0.50$$

The letter z is used to denote scores of the standard Normal distribution (in the same way that we might use X to denote ordinary scores), so the transformation formula can be written:

$$z = \frac{\text{value of } X - \text{mean of } X}{\text{standard deviation of } X}$$

The two z values just calculated, of $z = 0$ and $z = 0.50$, are marked on the standard Normal distribution in Figure 6.8(b), and we want the proportion of the total area under the curve represented by the shaded area between the z scores 0 and 0.5. Notice that the z score of 0.5 is not surprisingly one half of the standard Normal s.d. of 1, since 9 bpm is one half of the s.d. of 18 bpm. In other words the z score for any value of a Normally distributed variable X tells us *how many standard deviations that value of X is from its mean $\bar{x}$.*

As mentioned above, the proportions under the standard Normal curve have already been worked out and are contained in Table A1 in the Appendix at the end of this book. For convenience three rows of the standard Normal table are shown in Table 6.9.

Table 6.9: Three rows of the standard Normal or z table

z	0.00	0.01	0.02	0.03
0.4	0.1554	0.1591	0.1628	0.1664
0.5	**0.1915**	0.1950	0.1985	0.2019
0.6	0.2257	0.2291	0.2324	0.2357

MEMO

The standard Normal distribution has a mean of zero and an s.d. of 1, and the total area under the standard Normal curve is 1. Any original Normal distribution value can be converted into its equivalent standard Normal distribution value by subtracting from it the mean of the distribution and dividing by its s.d.

The first column, headed z, contains the value of z to the first decimal place. The entries in the main body of the table show the proportions of the total area under the standard Normal curve between the mean value of $z = 0$ and any given value of z. Our value is z is 0.50, so we need to look down the first column to locate the value of $z = 0.5$, and then to the second column, headed 0.00, because our second decimal place is 0. The table gives a value of 0.1915, and this is the proportion of the area under the standard Normal curve between $z = 0$ and $z = 0.50$. This is the *same* as the proportion of area under the original Normal heart rate curve between 123 bpm and 132 bpm.

In other words, 19.15% (just multiplying the proportion by 100 to give per cent) of the sample (and by inference, the population) have a heart rate of between 123 bpm and 132 bpm. The z transformation can be expressed algebraically. For any score X of the original distribution:

$$z = \frac{X - \bar{x}}{s}$$

where $\bar{x}$ is the sample mean and s the sample standard deviation.

Suppose the original question had been, "What proportion of the men in the sample had a heart rate greater than 132 bpm?". We can answer this very easily. We know that the total area under the standard Normal distribution is 1, and since all Normal curves are symmetric this means that the area in each *half* of the standard Normal curve must be 0.5. If the area between $z = 0$ (the middle) and $z = 0.5$ is 0.1915, then the area to the right of $z = 0.50$ must be $0.5 - 0.1915$, i.e. 0.3085. That is, 30.85% of the men had heart rates greater than 132 bpm.

We can use the z transformation in reverse, as it were. For example, assume that we want to know what heart rate value identifies those men whose heart rate puts them in the bottom 10%. This time we *start* by knowing what the proportion under the z curve is, i.e. 0.10 (the shaded area in Figure 6.9(a)). We need first to calculate the value of z which leaves an area of 0.10 in the left-hand tail, and then use the transformation in reverse to find the corresponding value of X in bpm in the original Normal curve.

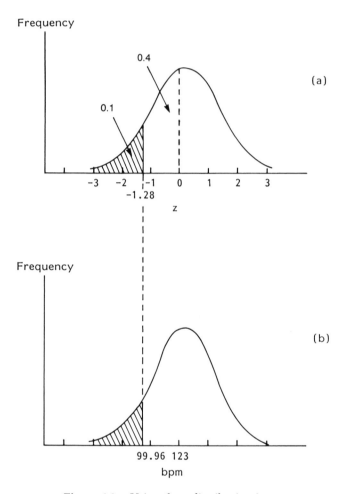

Figure 6.9: Using the z distribution in reverse

Remember that the z table in the Appendix gives the area under the z curve between z = 0 and the calculated value of z (most standard Normal tables are in this form, although some give the areas in the tail of the curve so care is necessary when using them). So the first step is to subtract the area in the tail from 0.50 (the total area under the left-hand half of the curve) to give us the area needed for the table. This area is thus 0.5 − 0.1 = 0.4, shown in Figure 6.9(a).

We now need to refer to the z table to find the value of z which has an area of 0.40 between it and z = 0, and for convenience the appropriate portion of the table is shown in Table 6.10. The value of 0.40 lies on the z = 1.2 row between the value 0.3997 (in the 0.08 column), and the value 0.4015 (in the 0.09 column).

Table 6.10: Portion of z table used to identify value of heart
rate (bpm) for lowest 10% of heart rate

z	0.07	0.08	0.09
1.1	0.3790	0.3810	0.3830
1.2	0.3980	**0.3997**	**0.4015**
1.3	0.4147	0.4162	0.4177

So the value we require is between $z = 1.28$ and 1.29. We could interpolate to get the exact value, but since 0.40 is very close to 0.3997 we will accept the value of $z = -1.28$ to be near enough. Notice that we have attached a *minus* sign to the z value because we are actually working in the left-hand area of the z curve where all values of z are *negative*. It is important to remember to do this! The calculated value of z is marked on Figure 6.9(a).

The final step is to use the transformation to get the required value of heart rate, X, in bpm. When we rearrange the expression z above we get:

$$\text{Value of } X = (z \times \text{standard deviation}) + \text{mean}$$

That is,

$$\text{Value of } X = (-1.28 \times 18) + 123 = -23.04 + 123 = 99.96 \text{ bpm}$$

In other words, the men with the lowest 10% of heart rates have a heart rate of 99.96 bpm or less. This value is shown in Figure 6.9(b).

We can use the standard Normal distribution to calculate the proportion of values less than, between, or greater than, any values of any Normal distribution we choose.

Finally, we want to use the z distribution to determine the *exact* areas under the Normal curve that lie within one, two, and three standard deviations of the mean, which we have previously expressed as "about" 68%, "about" 95%, and "about" 99%. To do this we need to refer to the complete table for the standard Normal distribution shown in Table A1 of the Appendix. Recall that this table gives the proportion of the total area under the curve between the mean ($z = 0$) and any given value of z. For $z = 1.00$ (i.e. one standard deviation) the table gives a value of 0.3413. Since the curve is symmetric this means that for one standard deviation *either* side of the mean, i.e. between $z = -1$ and $z = +1$, the proportion of the total area under the curve is $2 \times 0.3413 = 0.6826$. That is, 68.26% cent of the total area lies within one standard deviation of the mean. For $z = 2.00$, the value is 0.4772, so between $z = -2$ and $z = +2$ the proportion under the curve is $2 \times 0.4772 = 0.9544$. That is, 95.44% lies within two standard deviations of the mean. For $z = 3.00$, the value is 0.4987, so between $z = -3$ and $z = +3$ the proportion is $2 \times 0.4987 = 0.9974$. That is, 99.74% of the total area lies within three standard deviations from the mean. These values and areas are shown in Figure 6.10.

As a final point, the interquartile range and semi-interquartile range are often used as a measure of spread for metric variables particularly if a significant degree of *skew* is present in the sample distribution, and the presence of outliers may be thought liable as a consequence to produce a misleading value for the standard deviation.

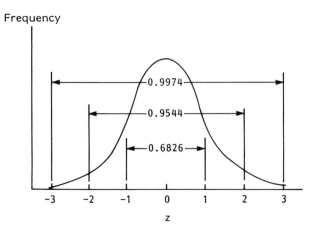

Figure 6.10: Exact proportions of the *standard* Normal curve

USING A COMPUTER TO GET MEASURES OF SPREAD

Unfortunately, I am not aware of any statistics package that calculates either the VR, the ID or the IQV. It would be possible no doubt to write small macros within SPSS and Minitab to overcome this problem, but I will leave this to you.

Using SPSS for measures of spread

In Chapter 5 we saw that SPSS can be used to produce summary measures of average and various percentile values with the commands:

   Statistics
     Summarize
       Frequencies

If the **Statistics . . .** box is then clicked, a dialogue box is displayed from which the user can choose a large number of optional measures, arranged in four sections: Percentile Values; Central Tendency; Dispersion; and Distribution.

In Figure 5.8, SPSS was used to produce measures of average and the three quartiles for the "before" and "after" Oswestry Mobility Scale scores from Table 3.6. If in addition we click on ⊠ **Std deviation**, ⊠ **Range**, ⊠ **Minimum** and ⊠ **Maximum**, we get the output shown in Figure 6.11.

Although I clicked on the "Std deviation" box to produce the value of 37.205 and 31.225 for the "before" and "after" standard deviation Oswestry scores in Figure 6.11, the fact that these scores are ordinal means that the standard deviation should *not* be used as a measure of spread (see Table 6.2 for a reminder of what is allowed). The interquartile range and semi-interquartile range are more appropriate measures for ordinal data, and can of course be calculated by subtracting the 25th percentile value from the 75th percentile value, for the IQR and dividing this result by 2 for the SIQR. If we wish to use the range this value is available in Figure 6.11.

```
SPSS for MS Windows Release 6.0

BEFORE

Mean 61.267     Median 66.500     Mode 10.000
Std dev 37.205    Range 105.000   Minimum 10.000
Maximum 115.000


Percentile  Value     Percentile  Value     Percentile  Value
25.00       22.500    50.00       66.500    75.00       96.250


Valid cases 30    Missing cases 31

AFTER

Mean 72.500     Median 72.500     Mode 40.000
Std dev 31.225    Range 117.000   Minimum 15.000
Maximum 132.000

* Multiple modes exist. The smallest value is shown.

Percentile  Value     Percentile  Value     Percentile  Value
25.00       47.500    50.00       72.500    75.00       100.000


Valid cases 30    Missing cases 31
```

Figure 6.11: Measures of spread produced by SPSS for the Oswestry Mobility Scale "before" scores of Table 3.6

Using Microsoft Excel for measures of spread

In Chapter 5 we used Excel to produce measures of average for the "before" Oswestry data (Figure 5.10). I have used exactly the same commands to produce descriptive statistics for the duration of the first ECT seizure using the raw data in Table 4.4. The output is shown in Figure 6.12. As you can see, in addition to measures of average, Excel also produces a number of measures of spread, e.g. standard deviation (SD); the minimum and maximum values (from which the range can be calculated); and the 25th and 75th percentiles (from which the interquartile range can be calculated).

Using Minitab for measures of spread

In Chapter 5 the following commands were used to get Minitab to produce the descriptive statistics shown in Figure 5.9:

Stat
  Basic Statistics
    Descriptive Statistics
      Select c1
        Select c2
          OK

```
Single descriptive statistics (column)

Range:   [Book1]Sheet1: First [2 − 41]

n:            40            Minimum:          15
Sum:          1403.0        25th Percentile:  25
Mean:         35.075        Median:           30
Variance:     218.1224      75th Percentile:  43.5
SD:           14.769        Maximum:          90
SE:           2.3352
Skewness:     1.4603        Mode:             30
Kurtosis:     3.7888
```

Figure 6.12:   Output from Excel of Descriptive Statistics for the duration of the first seizure (raw data in Table 4.4)

Minitab, unlike SPSS, does not offer a choice of output statistics, and in addition to the various measures of average, Minitab also calculates the standard deviation, the minimum and maximum values and Q1 and Q3 (see Figure 5.9), from which the IQR and SIQR can be calculated.

Using EPI Info for measures of spread

In Chapter 5 we saw that EPI can be used to produce summary descriptive statistics for the Oswestry data with the commands:

   Programs
      Analysis
         read osw.rec
            freq before after

The output produced is shown in Figure 5.10. This includes, in addition to the various measures of average, the minimum and maximum values, the 25th and 75th percentiles, and the standard deviation (see Figure 5.10 for details).

The Box Plot

The box plot is widely used to summarise data graphically in terms of its various centile values. It provides an illuminating visual display of the first and third quartile values, and of the median, as well as the maximum and minimum sample values.

To illustrate the use of a computer to produce a box plot, we can again use the data on the "before" and "after" Oswestry Mobility Scale scores, in Table 3.6.

Using SPSS to produce a box plot

The data are entered into columns c1 and c2 of the SPSS data sheet, and named "before" and "after" with the **Define Variables** command in the **Data** menu. The following command sequence will produce the box plot shown on Figure 6.13.

> **Graphs**
>> **Boxplot**
>>> **Simple**
>>>> ⊙ **Summaries of separate variables**
>>>>> **Define**
>>>>>> **Select before**
>>>>>>> **Select after**
>>>>>>>> **OK**

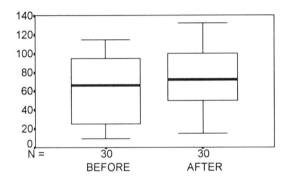

Figure 6.13:   SPSS box plot for Oswestry Mobility Scale "before" and "after" scores (raw data in Table 3.6)

The top and bottom of the box mark the first and third quartiles. The horizontal line inside the box marks the position of the median. The top of the vertical line coming out of the top of the box marks the largest sample value, the bottom of the line coming out of the bottom of the box the smallest value (although these may sometimes be replaced by the smallest and largest 5% and 95%, or other similar values).

Using Minitab to produce a box plot

To produce two box plots of the "before" and "after" Oswestry Mobility Scale scores on the same graph using Minitab, the before and after data are entered into the same column (named *Oswestry*), and a category value (1 = before, 2 = after) entered into the next column (named *When*). The following commands are then used:

> **Graph**
>> **Boxplot**
>>> **Select "Oswestry" as Graph 1 in Y column**
>>>> **Select "When" in X column**
>>>>> **Select "Oswestry" as Graph 2 in Y column**
>>>>>> **Select "When" in X column**

(Choose options in Data Display box, **Items 1** and **2**)
> **Frame**
> > **Multiple Graphs**
> > > **Overlay graphs on same page**
> > > **OK**
> > > **OK**

To copy the graph into another Windows document (e.g. this book), choose:

> **Edit**
> > **Copy graph**

This will copy the graph to the Windows Clipboard from where it can be pasted into another application (Figure 6.14). The SPSS and Minitab box plots are virtually identical, and provide very easy-to-see confirmation that the "before" Oswestry scores are lower and have wider spread than the "after" scores. I would encourage the use of box plots as an excellent visual device.

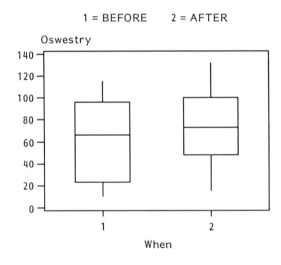

Figure 6.14: Box plot of Oswestry Mobility Scale "before" and "after" scores using Minitab

THE COEFFICIENT OF VARIATION

I want to conclude this chapter by mentioning briefly the coefficient of variation (cv). This is used occasionally to compare the spread in two distributions of the same variable which have different means and standard deviations, or when they are measured in different units, e.g. °C and °F, or g and kg. The coefficient of variation is found by dividing the standard deviation by the mean. The result is sometimes then multiplied by 100%:

$$cv = s/\bar{x} \ (\times \ 100\%)$$

Since the cv involves the mean and standard deviation it can only be used with metric data. Coefficients of variation less than 1 (or less than 100%) indicate a standard deviation "small" relative to the mean, i.e. relatively small spread; cv values greater than 1 (or 100%) indicate a standard deviation large relative to the mean, i.e. relatively large spread.

An Example from Practice

In a study of turnover and length of stay of staff in the NHS, the authors calculated the coefficients of variation for a number of different staff groups[5]. These are shown in Table 6.11. The authors defined turnover rate as the annual number of staff leaving divided by the average numbers of staff in post plus the number leaving.

Table 6.11: Coefficients of variation in the turnover and length of stay of staff in the NHS

STAFF GROUP	FULL-TIME	PART-TIME
Registered nurses and health visitors	33.1	47.0
Enrolled nurses	59.4	60.0
Midwives	47.9	66.2
Other nurses	42.8	48.7
Physiotherapists	44.9	76.4
Radiographers	60.3	109.6
Clerical officers	32.3	85.3

The authors commented, "The coefficient of variation is likely to be highest in the groups containing relatively small numbers of staff . . . and the Table bears this out. Generally the table suggests that . . . turnover rates for part-time staff varied to a greater degree than for full-time staff".

SUMMARY

In this chapter we have completed the discussion of the procedures with which we can describe sample data, by examining ways in which we can measure the spread or dispersion of the data around some central value.

With nominal data and non-numeric ordinal data we use the frequency-based measures, i.e. the variation ratio which measures the proportion of sample values not in the modal category; the index of diversity and the index of qualitative variation, both of which measure the degree of concentration of values into a few large categories. The advantage of the IQV is that it has a known maximum value (of 1) which is independent of the number of categories.

With numeric ordinal data and with skewed metric data we can use the range-based measures, i.e. the range, the interquartile range and the semi-interquartile range. The range is the simplest of these measures and is just the difference between the largest and smallest values. This measure is very sensitive to outliers, hence the use of the interquartile measures, which remove a percentage of the lowest and highest values.

With metric data (if reasonably symmetric) the best measure of spread is the standard deviation. The problem with the s.d. is that it is difficult to interpret in a common sense way.

EXERCISES

6.1 (a) Calculate the standard deviation for the scores in question 5.1.
   (b) Recalculate the standard deviation after: (i) adding 2 to each score; (ii) subtracting 2 from each score; (iii) multiplying each score by 2; (iv) dividing each score by 2.
   (c) Formulate some general rules governing the effect on the value of the standard deviation of adding, subtracting, multiplying and dividing scores by a constant number.
   (d) What effect do you think each of these operations would have on (i) the range; (ii) the interquartile range? Explain your reasoning.

6.2 Calculate and compare the variation ratio, the index of diversity and the index of qualitative variation for the hair length and colour data shown in Table 5.13.

6.3 Table 6.12 records the number of successful suicides in 1980 in Copenhagen by diagnosis and sex[6]. Calculate the VR, the ID and the IQV for males and females and compare differences in the spreads of the data.

Table 6.12:   Number of successful suicides by sex and diagnosis in Copenhagen in 1980

DIAGNOSIS	MALE	FEMALE
Schizophrenia	2	4
Affective psychosis	1	7
Psychogenic psychosis	0	2
Neurosis	2	7
Personality disorder	12	8
Alcoholism	16	12
Drug addiction	2	1
Other	2	7
Not mentally ill	13	14

6.4 Table 6.13 records the adequacy of information, in the opinion of nurses, given to 125 breast cancer patients on internal and external prostheses before and during hospitalisation (in per cent)[2]. Calculate the VR, ID and IQV for all four situations and comment on what is revealed. Would it be possible to calculate the interquartile ranges? Explain your answer.

6.5 In question 5.5 you were asked to calculate the median and modal Waterlow pressure-sore risk scores for the 40 patients whose data appear in Table 3.14. Calculate the interquartile range and semi-interquartile range scores for the same data and interpret your results. What do you think about using the standard deviation as a measure of spread for these data?

Table 6.13: The adequacy of information, in the opinion of nurses, given to 125 breast cancer patients on internal and external prostheses before and during hospitalisation (percentages)

(a) BEFORE HOSPITALISATION

	None	Little	Not known	Much	Very much
Internal prosthesis	23	42	29	5	1
External prosthesis	7	46	22	23	2

(b) DURING HOSPITALISATION

	None	Little	Not known	Much	Very much
Internal prosthesis	36	46	9	8	1
External prosthesis	15	32	5	44	4

6.6 In question 5.9 you were asked to calculate the three measures of location for the data on the number of smoking co-workers of subjects in a study of smoking and heart disease (original data in Table 4.8). For this same data, calculate the standard deviation and interpret your result.

6.7 In question 5.6 you were asked to calculate the mean and median for the time spent waiting by 60 patients to see their GP (raw data in Table 4.9).

 (a) Calculate the interquartile range and semi-interquartile range for this same data. Comment on what your results mean.
 (b) Calculate the standard deviation waiting time. Comment on your result and compare with the results in (a).

6.8 In a study of the relationship between diet and coronary heart disease, researchers gathered the data in Table 6.14, which records the consumption

Table 6.14: Consumption of wine, beer and spirits ethanol (annual litres per head) in 1988, in 21 countries

	WINE ETHANOL	BEER ETHANOL	SPIRITS ETHANOL
Australia	2.5	5.5	1.2
Austria	3.9	5.9	1.5
Belgium/Luxembourg	2.9	4.4	2.2
Canada	2.4	4.2	2.5
Denmark	2.9	5.1	1.5
Finland	0.8	2.9	3.2
France	9.1	1.6	2.4
Iceland	0.8	0.3	2.4
Ireland	0.7	4.7	1.7
Israel	0.6	0.6	3.1
Italy	7.9	1.3	1.0
Japan	1.5	2.2	2.3
Holland	1.8	4.2	2.1
New Zealand	1.9	6.1	1.6
Norway	0.8	2.3	1.3
Spain	6.5	2.7	3.0
Sweden	1.6	2.0	2.0
Switzerland	5.8	3.3	2.0
UK	1.3	5.5	1.7
USA	1.2	3.6	2.4
Germany	2.7	5.8	2.2

of wine, beer and spirits ethanol (annual litres per head) in 1988, in 21 countries[7]. Use the methods of descriptive statistics discussed in Chapters 4, 5 and 6 to describe these data as thoroughly as possible.

6.9 In a study of the use of diagnosis-related groups (DRGs) in the NHS[8], the authors compared the coefficients of variation for lengths of stay in the UK, Canada and the Netherlands. The data are shown in Table 6.15. Comment on what is revealed by the data.

Table 6.15: Coefficients of variation in lengths of hospital stay from a study on the use of DRGs in the NHS

COEFFICIENT OF VARIATION	NUMBER OF DRGs		
	UK	Canada	Netherlands
0.20–0.39	11	13	57
0.40–0.59	56	63	125
0.60–0.79	96	127	140
0.80–0.99	143	144	63
1.00–1.19	109	59	15
1.20–1.39	38	7	6
1.40+	6	7	1

6.10 In an investigation into pregnancy dating using ultrasound[9], the researchers determined the sample mean and sample s.d. maternal height (cm) for the mothers of (i) 162.5 cm and 6.3 cm for the English/European mothers; and (ii) 158.9 cm and 5.9 cm for the Indian or Pakistani mothers (Table 6.16).

Table 6.16: Parity by ethnic group for a large sample of mothers in an investigation of pregnancy dating by ultrasound (percentages)

PARITY	ENGLISH/ EUROPEAN	INDIAN OR PAKISTANI
0	44.4	30.8
1	35.3	29.6
2	13.7	18.9
3	4.5	9.8
≥ 4	2.1	10.8

(a) How reasonable would it be to assume that maternal height is Normally distributed?

(b) Assuming Normal distributions, sketch the frequency curves for each of the two ethnic groups.

(c) Use the *standard* Normal distribution to determine for each of the two ethnic groups: (i) the percentage of mothers who are more than 165 cm tall; (ii) the percentage of mothers who are less than 150 cm tall; (iii) the percentage of mothers who are between 165 cm and 150 cm; (iv) the percentage of mothers who are more than 150 cm tall.

(d) Ten per cent of the mothers in each of the two ethnic groups are: (i) less than how tall? (ii) taller than what height?

REFERENCES

1. Vasiliadou, A. *et al.* (1995) Occupational low-back pain in a Greek hospital. *Journal of Advanced Nursing*, **21**, 125–30.
2. Suominen, T. (1993) How do nurses assess information received by breast cancer patients? *Journal of Advanced Nursing*, **18**, 64–8.
3. Zoltie, N. and de Dombal, F. T. (1993) The hit and miss of ISS and TRISS. *BMJ*, **307**, 906–7.
4. Bhambhani, Y. (1994) The Baltimore Therapeutic Work Simulator: biomechanical and physiological norms for three attachments in healthy men. *American Journal of Occupational Therapy*, **48**, 22–4.
5. Gray, A. M. and Phillips, V. L. (1994) Turnover, age and length of service: a comparison of nurses and other staff in the National Health Service. *Journal of Advanced Nursing*, **19**, 819–27.
6. Nordentoft, M. *et al.* (1993) High mortality by natural and unnatural causes: a ten-year follow-up study of patients admitted to a poisoning treatment centre after suicide attempts. *BMJ*, **306**, 1637–41.
7. Criqui, M. H., *et al.* (1994) Does diet explain the French paradox? *The Lancet*, **344**, 1719–20.
8. Sanderson, H. *et al.* (1986) Using diagnosis-related groups in the NHS. *Community Medicine*, **8**, 37–46.
9. Wilcox, M. *et al.* (1994) Birth weight from pregnancies dated by ultrasonography in a multicultural British population. *BMJ*, **307**, 588–91.

7

GETTING THE DATA

Chapters 2 to 6 described a number of ways of exploring sample data. We can construct frequency distributions and draw charts to get the broad picture and/or calculate measure of average and of spread if we want to be more exact. But how do we get the data in the first place? That's the issue I will deal with in this chapter.

SAMPLING AND SAMPLES: THE GOOD, THE BAD AND THE UGLY

In simple terms a *sample* is a collection of individuals or items taken from a population. As we noted in Chapter 1, a *population* contains *every* person or item in an entire group (however we define that group). Thus a population might be, for example, *every* person with a history of coronary heart disease in San Francisco, or *every* female in the UK, aged from 18 to 30, who is HIV+, or *every* prescription written last year in a large general practice, or *every* level of diastolic blood pressure it is possible to measure, and so on*. The important thing is that, whatever the population, it is defined to include *every* possible eligible member.

But why do we study samples and not populations? Here are six reasons:

- Because populations are usually too big for us to have either the time or the resources to locate and study every single member of them. Imagine the difficulty of finding and interviewing everybody in Britain who had taken at least one over-the-counter analgesic tablet in the last seven days. In any case, as we have seen, some populations are infinite in size.
- Because we need to restrict the size of the sample for ethical or safety reasons (giving a new drug its first use with human volunteers, for example).

---

* Remembering what we said about continuous variables in Chapter 2, this last population, like many others, has an infinite number of "items" or values in it.

- Because of experimental reasons, e.g. not being able to find very many suitably matched individuals for the subjects in a case–control study (more on this later).
- Because the members of the population are very difficult to identify (for example, individuals carrying a virus which produces few noticeable signs or symptoms until at a very late stage).
- Because a population is too transient or too volatile to measure (for example, all injecting drug abusers in Edinburgh).
- Because the sampling process itself is destructive (you can't take too many biopsy samples from a patient, or test many new wheelchairs to destruction before signing a contract to buy 50!).

When we have a sample drawn from some population our hope is that it is *representative* of that population, i.e. that all the features of the population are accurately reflected in the sample. The more closely the sample resembles the population, the more reliable will be any conclusions about the population based on the sample. Given all this then, how do we actually go about taking a sample so that it is as representative of the population as possible?

MEMO

A sample is a collection of persons or items taken from a population. A population includes *every* eligible member however defined. There are a number of reasons why we study samples and not populations, not least because populations are usually too big or too difficult to identify.

The Simple Random Sample

Badly chosen samples run the risk of being *biased*. For example, suppose we were investigating what patients thought of the appointment system to see a GP at a local health centre. We decide to interview *every* patient who sees the GP during one particular week, thinking that this would give us a representative sample. In fact, this method would produce a biased sample because, for one thing, it would omit all of those patients who had tried but been unable to make an appointment. This sort of bias is called *selection bias*.

A better approach would be to get a list of *all* the patients on the GP's list (which would constitute the population in this example), write each name on a piece of paper, put them all in a hat and then draw out say 100 of them whom we could then interview. In this way each patient has an equal chance of being picked for inclusion in the sample, and the sample will now be representative of the population of patients. This approach, where every individual or item has an equal chance of being picked for inclusion in the sample, is called *random sampling*.

Unfortunately, even with smallish populations, writing out this many pieces of paper would be very tedious. Much better would be to use a table of *random numbers*, such as that in Table A2 in the Appendix at the end of this book. We would first give each patient on the list a number and then stick a pin somewhere in the random number table. From this starting point we would move up or down or across the page, writing down the successive random numbers. Suppose that the first few numbers we got were:

    20428   51660   91083   17299   39198 . . .

If we assume that there are 9000 patients registered with the practice altogether, then the first patient in the sample would be the one numbered 428, the second number 1660, the third 1083, the fourth 7299, and so on until there were 100 patients in the sample. Of course any number greater than 9000 is ignored.

The essential starting point for such a procedure is a numbered list of the name and address of every member of the population in question. Such a list is known as a *sampling frame*, but for many populations producing such a list may be very difficult if not impossible (we will return to this problem shortly).

MEMO

Simple random sampling is when every member of the population has an equal chance of being picked for inclusion in the sample.

Systematic Random Sampling

One of the problems with the above approach is that the table of random numbers may produce duplicated values, so we will have to trawl through our completed sample to check that no individual is included more than once. Besides which, with large samples picking out and writing down the random numbers can be time-consuming and prone to error.

"It's amazing; he gets a 100% return on all his questionnaires."

A variation on the simple random sample approach which overcomes both of these difficulties is *systematic* random sampling. In this we start by dividing the number in the population by the required sample size to produce the *sampling fraction*. In the above example the total number of patients on the list is 9000. With a sample of 100 this gives a sampling fraction of 9000/100 = 90. We then choose at random a start number between 0 and 90, say 72, and then pick every 90th person in the sampling frame. This means picking the 72nd, 162nd, 252nd, etc. until we have 100 people for the sample, which should just be possible. This is a little easier and quicker than the previous approach, but if the sampling frame has any sort of underlying pattern, i.e. is not truly random, then the sample produced may be correspondingly biased.

Stratified Random Sampling

One drawback with the simple random sample is that there is no guarantee that it will contain the same proportions of, say, males and females, or age groups, or of other *strata* there are in the population. If we want to ensure that the relevant strata are present in the sample in the same proportions as in the population, then we first divide the sampling frame into the relevant strata, then take a simple random sample from each strata. These separate samples are then combined to produce a *stratified random sample*.

For example, we may want to conduct a survey among nursing staff in a District Health Authority of their attitude to a major restructuring proposal. Suppose that we know that the percentages of the various grades of nursing staff are as follows:

Nursing managers and tutors:	3%
Registered nurses and health visitors:	58%
Enrolled nurses:	11%
Midwives:	7%
Other nursing staff:	21%

If we have decided on a sample size of 1000, then to ensure that all grades are properly represented in the sample, we would need to take a separate random sample from each group of 30 nursing managers and tutors, 580 registered nurses and health visitors, 110 enrolled nurses, 70 midwives, and 210 other nursing staff. In this way, the proportion of each grade of nurse will be the same in the sample as in the population.

MEMO

Most random sampling procedures require as a prerequisite a list, known as the sampling frame, of every member of the population.

The Multistage Random Sample

A disadvantage of all the above approaches is that a sampling frame is a prerequisite. For many populations this may be difficult or impossible to obtain. One way around this is to use the *multistage random sample*. Suppose for example that we wanted a sample of all patients in a region of the country who were admitted to hospitals as a consequence of a road traffic accident. We would first obtain a list of all the hospitals in the region and take a random sample of these (the first stage). From each of these (much fewer) hospitals we could more easily obtain a sample frame. The second stage is then to take our random sample from the patients on this sample frame.

A variant of this approach is to include every member of the last stage in the sample. This is known as *cluster* sampling.

Despite the virtues of random sampling for achieving samples that are representative of their parent populations, in many health care applications such random sampling is not possible. This may be because we can't properly define or identify the population we wish to study; or because, even if we can identify it, we can't, for ethical, professional or practical reasons, gain access to it. How for example can we access the population of women who have had an abortion, or define the population of all those with high blood pressure, many of whom may not even be aware of their condition?

In practice the usual raw material is a sample of *presenting* patients (sometimes known as a *sample of convenience*), or perhaps a sample of volunteers (students or colleagues, or self-selected members of the public). For example, a sample of 100 consecutive patients with a particular gastrointestinal disorder attending an out-patient clinic, or a sample of patients with asbestosis, or a sample of volunteers taking some new drug, and so on. Such samples can in no way be described as being random samples from some wider more general population, and it would be extremely unwise to draw any conclusions about *any* population other than the very restricted one from which such samples are drawn. However, we can hope that, at the very least, some useful insights into *similar* populations can be obtained, or that useful conclusions can be drawn when two such samples are compared (e.g. the efficacy of a new intervention).

Accepting these various limitations on the representativeness of many samples in practice, we shall now consider a number of ways in which we can use samples to investigate problems of interest.

STUDY DESIGN

As has been pointed out elsewhere[1], there are several ways in which we can classify the procedures for carrying out investigations in the field of health care. For example:

- as observational versus experimental studies[2]
- as prospective versus retrospective studies
- as longitudinal versus cross-sectional studies[3].

For our purposes it is convenient to use the first of these classifications. The essential difference between observational and experimental studies is that in the former we simply observe what's going on in different groups of individuals without having any control over who is in which group, whereas in the latter we make some sort of intervention, for example allocating some individuals to one group and some to another. This distinction should be clearer after we have looked at some examples. Observational studies include cohort studies, case–control studies and cross-sectional studies. Experimental studies include clinical trials.

MEMO

Study designs can be broadly divided into observational or experimental. In experimental designs the investigator has control over who is allocated to which group.

FROM HERE TO ETERNITY—COHORT STUDIES

The cohort approach is usually concerned either with the cause (aetiology) of a disease, or with the prognosis of those already suffering from some disease. It is often called a *longitudinal* or *prospective* study because it follows a group of individuals over some period of time, until some *end point*, usually either until death or the development of the disease of interest is reached. A cohort may be time-based, such as a sample of all those born in 1990, or discharged from a certain hospital in 1995, or it may consist of individuals associated with a certain occupation, such as children whose fathers work in the nuclear power industry.

At the end of the study (usually several years, i.e. until a sufficient number of the subjects have died or experienced further illness) the exposure or non-exposure of those involved to some risk factor is determined. *Risk* is defined as the probability that some event will occur within some given time period. For example, one famous cohort study[4] included all doctors in the UK and identified those who smoked and those who didn't. The end point of this study was death, which was found to occur significantly earlier among those doctors in the cohort who smoked. The nature of a cohort study is shown schematically in Figure 7.1

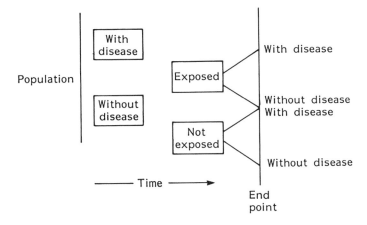

Figure 7.1:   Schematic of cohort study

The cohort approach is not well suited to the investigation of rare diseases, since this would require very large samples extending over long periods of time, before a sufficiently large number of subjects had died or reached some other defined end point. Large samples will consume correspondingly large resources of time and personnel in follow-up, and results may be a long time coming. Moreover, the longer the study the more subjects will be lost, for example because they move out of the area, otherwise lose contact, or stop responding or complying, or die from causes not connected to the study (e.g. in road traffic accidents). These losses decrease the accuracy of the study and increase the possibility of *bias*, because some of the losses may be due to an unrecognised connection with the outcome being measured in the study.

Relative Risk

We can compare the likelihood of developing a disease in those exposed to a risk factor with the likelihood in those not exposed—for example the risk of developing lung cancer in those who smoke compared with the risk in those who don't smoke. We do this by expressing the exposed risk as a ratio of the non-exposed risk. This ratio is called the *risk ratio* or more commonly the *relative risk*. To see how this is calculated, we first formulate the cohort study in tabular form as in Table 7.1.

Table 7.1:   Tabular formulation of cohort study

		WITH DISEASE	
		Yes	No
EXPOSED TO RISK	Yes	a	b
	No	c	d

Total number in the cohort, $n = a + b + c + d$
Number with disease $= a + c$
Number without disease $= b + d$
Number exposed to risk $= a + b$
Number not exposed to risk $= c + d$

The *absolute* risk of developing the disease for those exposed to the risk factor is $a/(a + b)$. The *absolute* risk of developing the disease for those not exposed to the risk factor is $c/(c + d)$. The ratio of these two gives us the relative risk and is written:

$$\text{Relative risk, or } RR = \frac{\dfrac{a}{(a + b)}}{\dfrac{c}{(c + d)}}$$

Relative risk measures the *increased chance* of morbidity or mortality among those exposed to the risk compared with those not exposed.

An Example from Practice

Danish researchers investigating the influence of alcohol on mortality studied a cohort of 1394 men[5]. Of these, 1183 men had been identified as light drinkers (between one and six beverages a week), and 211 men as heavy drinkers (70 or more beverages a week). A beverage was defined as 9–13 g of alcohol and equals one bottle of beer, one glass of wine or one measure of spirits. The study was started between 1976 and 1978 and the cohort followed up to 1988 or death. At this time the total mortality among the light drinkers was 252 men (931 still alive) and among the heavy drinkers 66 men (145 still alive). These data are arranged in Table 7.2.

Table 7.2:   Cohort study investigating the relationship between alcohol consumption and mortality

		DIED		TOTALS
		Yes	No	
EXPOSED TO RISK	Yes (≥ 70)	66	145	211
	No (1–6)	252	931	1183

The increased risk of mortality between the light and heavy drinkers can be measured by calculating the relative risk. Substituting into the above formula gives the relative risk as:

$$RR = \frac{\dfrac{66}{(66 + 145)}}{\dfrac{252}{(252 + 931)}} = \frac{0.3127}{0.2130} = 1.468$$

In other words, a heavy drinker was nearly one and a half times more likely to have died during the follow-up period than a light drinker. A relative risk of 1 means that there is no increased risk from exposure to the "risk" factor. A relative risk of less than 1 means that exposure to the "risk" factor is protective or beneficial.

Relative risk is an extremely good measure of the closeness of the association between a risk factor and a disease. It is unlikely to be affected by unidentified confounding variables (see below) since their influence will be felt in both the numerator and denominator of the expression for RR and therefore will cancel out.

Using a Computer to Calculate Relative Risk

Calculating the relative risk is easy once the values of a, b, c and d in Table 7.1 have been determined. However, the EPI package will calculate the RR if necessary.

Using EPI Info to calculate RR

With EPI, use the command sequence:

> **Program**
> > **Statcalc**
> > > **Tables**

The screen then displays a blank 2 × 2 table, labelled Exposure, + or –, down the left-hand side (the rows) and Disease, + or –, across the top (columns) with the cursor in the top left-hand cell. The values of a, b, c and d should then be typed in one after the other, following each entry by the return key. When all four values have been entered a further press of the return key will produce the value of the relative risk (along with a lot more information). The final screen looks like Figure 7.2.

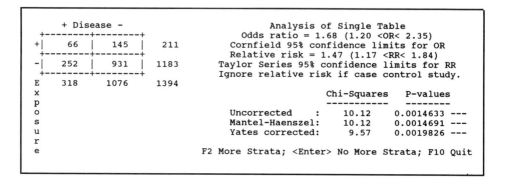

Figure 7.2: Relative risk produced by EPI for the alcohol mortality data

It is not possible to calculate the relative risk directly with either Minitab, SPSS or Microsoft Excel.

BACK TO THE FUTURE: CASE-CONTROL STUDIES

Whereas cohort or prospective studies go *forward* from the risk to the end point, case–control or *retrospective* studies start at the end point and go backwards in time to try to identify risk factors which the subjects (known as the *cases*) might have been exposed to in the past. Any findings are compared with those obtained from a comparison group (called the *controls*). The nature of a case–control study can be shown schematically as in Figure 7.3.

For example, we might select a sample of babies who had died from sudden infant death syndrome (SIDS) as the cases, and a sample of healthy babies of the same age as the controls (see below for an example). We would then try to discover what factors were different between the two groups, usually based on some hypothesis we had, for example whether the parents smoked, or whether the baby shared a bed with the parents, or the type of mattress used, and so on. If we found that more of the cases (SIDS babies) had parents who smoked, than did the controls, then we might conclude that smoking by parents increases the risk of SIDS.

The most difficult part of a case–control study is choosing the controls. Ideally the controls should be as similar to the cases as possible without having the disease, i.e. they should be a sample from the same population as the cases. For example, if the cases are hospital patients, the controls might also be patients in the same hospital but admitted for a different condition.

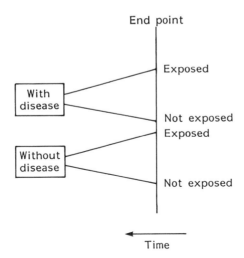

Figure 7.3:   Schematic of case–control study

A major problem in the selection of appropriate controls is the presence of *confounding* variables. To illustrate the idea of confounding, suppose we are studying a suspected relationship between people who have migraine and those who have a myocardial infarction (MI). Table 7.3(a) shows a group of 12 cases (people who have had an MI) and 12 controls who have been selected as being as similar to the cases as possible. Those who suffer from migraine have an "M" after their name, those who don't have "NM". There are 10 of the cases without migraine and two with, while there are five of the controls without migraine and seven with. At first sight we might conclude that having migraine appears to protect against MI.

Notice now that an "A" after the name means that these individuals took aspirin. Since those with migraine tend to take more aspirin than those who don't, and aspirin is a possible defence against MI, we would expect that those with migraine might suffer fewer MIs than those without migraine. In this case the taking of aspiring is a *confounding variable* and we should make allowances for this by excluding from the study all those who take aspirin.

When we do this (Table 7.3(b)) we see that there are now, among the cases, one subject with migraine and seven without, while among the controls there is one with migraine and five without. We would be far less inclined to draw the conclusion that suffering from migraine protects against MI, having excluded the confounding variable. Notice that we don't have to have equal numbers of subjects in the case and control groups (unless we are using a matched design—see below). In fact, control groups are often considerably larger than the cases, and more than one control group may be selected.

To be a confounding variable, a factor must be associated with *both* the exposure *and* the outcome of interest. Sex and age are common confounding variables and case–control studies often correct for these by choosing controls so that the distributions by age and sex are similar to those in the cases. For example, if 10% of the cases are males aged between 20 and 29 then the same percentage of men in that age group is selected as controls (this is known as *frequency matching*).

Table 7.3: Case–control study into the relationship between migraine and myocardial infarction. (M = migraine sufferer; NM = not a migraine sufferer)

(a) ORIGINAL DATA WITHOUT CONTROLLING FOR THOSE WHO TAKE ASPIRIN

Cases (myocardial infarction)	*Controls* (no myocardial infarction)
John (NM)	Rod (M, A)
Jack (NM, A)	Jess (NM)
Sue (NM)	Dick (M, A)
Mary (M, A)	Tom (NM)
Harry (NM)	Harry (NM)
Paul (NM)	Bill (M, A)
Jane (M, A)	Frank (M, A)
Liz (NM)	Mark (M, A)
Ray (NM)	Alice (NM)
Felicity (NM, A)	Bob (M, A)
Dan (NM)	Tom (NM)
Piers (M)	Cressida (M)
M = 2	M = 7
NM = 10	NM = 5

(b) EXCLUDING THOSE WHO TAKE ASPIRIN

Cases (myocardial infarction)	*Controls* (no myocardial infarction)
John (NM)	Jess (NM)
Sue (NM)	Tom (NM)
Harry (NM)	Harry (NM)
Paul (NM)	Alice (NM)
Liz (NM)	Tom (NM)
Ray (NM)	Cressida (M)
Dan (NM)	
Piers (M)	
M = 1	M = 1
NM = 7	NM = 5

MEMO

A major problem with case–control studies are confounding variables. These are variables that are associated with *both* the disease *and* the risk factor, whose presence may distort or disguise the perceived relationship.

An alternative procedure is to match each *individual* in the cases group with an individual in the control group by age and sex (or whatever the confounding variables are thought to be). Such arrangements give rise to *matched* case–control studies. In general, the number of confounding variables controlled can vary from one or two to many. For example, in a study[6] of the relationship between blood pressure and mortality in elderly people the authors felt it necessary to control for 17 variables (age; sex; history of myocardial infarction, of stroke, of angina, of cancer; use of antihypertensives, of loop diuretics, of digoxin, of hypoglycaemics; need for help with activities of daily living; problems with physical function; low activity level; overweight; current smoking, past smoking; and current use of alcohol). The kitchen sink was not included.

Apart from the problems of choosing appropriate controls, the other problem with case–control studies is the fact that such retrospective studies rely on peoples' memories, which can lead to bias (those with the disease can often recall events which may relate to their illness better than those without), or on the accuracy of historic data, which may be unreliable. One major advantage of the case–control study over the cohort study is that results are available much earlier, but the problems with choosing suitable controls means that cohort studies are generally preferred in terms of reliability of results. The case–control study can be formulated in tabular form. Table 7.4(a) shows the formulation for an unmatched study, Table 7.4(b) that for a matched study.

Table 7.4:   Tabular formulations of two types of case–control studies

(a) UNMATCHED DESIGN

		Cases (with disease)	Controls (without disease)
Exposed to risk factor	Yes	a	
	No	c	b d

(b) MATCHED DESIGN

		CONTROLS		Totals
		Exposed	Not exposed	
	Exposed	e	f	a
CASES	Not exposed	g	h	c
	Totals	b	d	

Number of cases = a + c
Number of controls = b + d
Number exposed to risk factor = a + b
Number not exposed to risk factor = c + d

An Example from Practice

Chinese researchers used an unmatched case–control study to examine the relationship between passive smoking at work and coronary heart disease (CHD) in Chinese women who had never smoked[7]. The cases were 59 Chinese women with CHD, and the controls were 126 Chinese women; both groups worked full-time and none of the subjects smoked. The values are shown in Table 7.5.

The results showed that among women exposed to passive smoking at work the chance of them developing CHD was more than twice that for women not so exposed. Moreover the likelihood of developing CHD increased as the number of smokers at work increased.

Table 7.5: Unmatched case–control study into the effect of passive smoking at work on coronary heart disease (CHD) among Chinese women workers

PASSIVE SMOKING AT WORK	CASES (with CHD)	CONTROLS (without CHD)
Yes	33	43
No	26	83
Totals	59	126

The Odds Ratio (OR)

In a cohort study we can use the relative risk (the ratio of two absolute risks) to measure the strength of any association between a risk factor and a disease, as well as to quantify the increased chances of getting a disease through exposure to a risk factor. However, in a case–control study we can't determine absolute risk because no group is followed forward in time, and so we can't calculate the relative risk directly.

Instead we derive an approximate measure of the relative risk by calculating what is called the *odds ratio*. The odds of an event is defined as the ratio of the number of persons experiencing the event to the number of persons not experiencing the event. Provided that the incidence of the disease in question in the population is small, and the cases and controls are random samples from their respective populations, then the odds ratio can be taken as a reasonable approximation to relative risk.

Thus, in the unmatched case in Table 7.4(a), the odds of those with the disease (the cases) having been exposed to the risk factor is a/c. For the controls (those without the disease), the odds of having been exposed is b/d. The odds ratio is the ratio of these two odds, that is:

$$OR = \frac{a/c}{b/d}$$

In the passive smoking example above:

$a/c = 33/26 = 1.2692 \qquad b/d = 43/83 = 0.5181$

Thus

$OR = 1.2692/0.5181 = 2.4498$

Thus the odds of a female worker exposed to passive smoking at work developing CHD is two and a half times the odds of a female worker not exposed to passive smoking at work developing CHD. Interpreting the odds ratio as an estimate of relative risk means that a working female passive smoker has nearly two and a half times the chance of getting CHD as a working female non-passive smoker.

In the matched case–control study, we do not have enough information to distinguish between the degree of exposure to risk of the cases and the controls. Referring to Table 7.4(b), the odds ratio in this case is:

$$OR = f/g$$

An Example from Practice

Researchers used two matched case–control designs each lasting 18 months, to investigate a possible link between deaths of babies from SIDS (sudden infant death syndrome) and bottle feeding[8]. For each of the 98 babies who died of SIDS, two babies were chosen as controls. The 196 controls were chosen from the same health visitor list as the SIDS babies and were matched for age, time of death and locality. Parents were asked about feeding method and babies were categorised as fully breast fed, mixed breast and bottle fed, and fully bottle fed. Potential confounding variables were identified as employment status, maternal smoking, sleeping position, and length of gestation.

The crude odds ratios for the combined results were, compared with fully breast fed babies, 1.5 for mixed breast and bottle, and 3.1 for fully bottle fed. After adjusting for sleeping position, maternal smoking, gestation, and employment status, these odds ratios were reduced to 1.2 and 1.8 respectively. The authors concluded that there was no clear evidence that breast feeding protects against sudden infant death syndrome, although bottle fed babies are more likely to have mothers who smoke, to be born pre-term, and to have come from poorer families, factors which are in themselves linked with sudden infant death syndrome.

Using a Computer to Calculate the Odds Ratio

As with the relative risk, calculation of the odds ratio is unlikely to require the power of a computer. The EPI program does not distinguish between the layout of the cohort study (Table 7.1) and that of the case–control study (Table 7.4) and simultaneously provides values for both the relative risk and the odds ratio, as can be seen in Figure 7.2.

Figure 7.4 shows the output from EPI for the unmatched case–control study into the effects of passive smoking by Chinese women workers (data in Table 7.5).

CROSS-SECTIONAL STUDIES

Unlike the cohort and case–control studies described above, the cross-sectional study looks at a particular population at some moment in time (or over a short

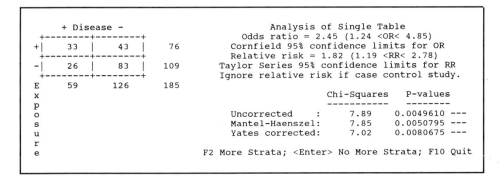

Figure 7.4:   Odds ratio from EPI for data on passive smoking and coronary heart disease

time period) rather than looking back or forward over a more extended period. The sample studied might be from a defined population, or it might consist of a sample presenting with some particular disease or condition.

Cross-sectional studies may be essentially descriptive in nature—for example finding the proportion of patients in a hospital with pressure sores, or the public's awareness of the symptoms of a particular disease—in which case they may then be referred to as *surveys*. Or they may be undertaken to study possible relationships between diseases and risk factors—for example, air pollution and hospital respiratory-problem admissions during a poor air-quality day in winter—in which case they serve as alternatives to the cohort or case–control approaches.

One very common application of the cross-sectional study is the measurement of the *prevalence* of some disease or condition (prevalence is the number of persons with a disease or condition at a particular point in time divided by the total population, and is usually expressed as a percentage, or per thousand or hundred thousand of the population).

Cross-sectional studies often gather their data using questionnaires, completed either face to face (by the subject or by the interviewer) or sent through the post. In the latter case, the response rate is an acknowledged weakness of this approach, since those who do not respond often differ in characteristics of interest from those who do respond. This can lead to bias in the results.

More detailed information on designing and implementing good questionnaires can be found elsewhere[9], but four golden rules are:

- Try to use an established questionnaire.
- Keep the format as simple as possible.
- Use as few questions as possible.
- ALWAYS pilot the questionnaire.

MEMO

A major problem with questionnaires (used extensively in cross-sectional studies) is lack of response, which can seriously bias results.

An Example from Practice

In 1994, researchers[10] monitoring young people's knowledge and experience of illicit drugs between 1969 and 1994 asked 392 school children aged 14 to 15 to complete anonymously the same questionnaire previously used in 1969, 1974, 1979, 1984 and 1989. There were seven questions in the questionnaire and all the children present in the school on the survey day completed at least some part of the questionnaire, although 12% were absent, a possible source of bias in the results.

The survey showed a "dramatic increase between 1969 and 1994 in young people's experience of and contact with illicit drugs". The proportion of children who knew someone taking drugs more than quadrupled from 15% to 65%, and the proportion who had been offered drugs increased ninefold from 5% to 45%. Both of these proportions more than doubled between 1989 and 1994.

Cohort, case–control and cross-sectional studies are compared schematically in Figure 7.5.

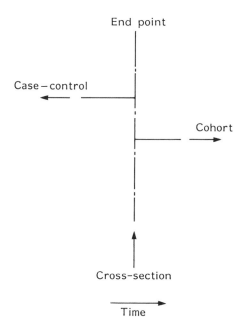

Figure 7.5: Schematic comparison of three study designs

EXPERIMENTAL STUDIES—THE CLINICAL TRIAL

The literature on the subject of clinical trials is vast and we will restrict ourselves here to only a very brief outline of the most important points. Clinical trials are principally used to assess the efficacy of new treatments or interventions (e.g. new drugs) compared with existing treatments. The commonest approach is known as the *parallel design*, in which the new treatment is given to one group (the *treatment* or *test* group) and at the same time the conventional treatment (or a placebo) is given to a second group (the *control* group).

A less frequently used approach is the *crossover* design in which one half of the subjects (chosen at random) are given one treatment and the other half the second treatment. After a short *washout* period to prevent carry-over effects, the two groups then swop treatments. The crossover design has the advantage that smaller sample sizes are needed, but suffers a number of limitations. For example it is obviously not suitable for conditions that can be cured, since this leads to the possibility that some subjects will drop out (cured) after the first stage. Dropout after the first stage for other reasons is, anyway, another potential problem.

The structure of this type of study is captured by the following quotation, taken from an investigation into the effects on memory of two drugs (biperiden and amantadine) used in the treatment of chronic schizophrenia[11]: "The patients were then randomly assigned in a double-blind manner so that half received biperiden and the other half amantadine. This first treatment phase continued for two weeks after which the patients were switched to identical placebo for one week. Each patient then received the alternative active drug for a further two week period". (For the meaning of double-blind, see below.) The patients were assessed using a number of tests of memory at the beginning of the first treatment phase, at the end of the first treatment phase, at the end of the placebo phase, and at the end of the second treatment phase. As a matter of interest, the authors concluded that "Biperiden treatment was associated with significantly poorer memory performance than amantadine treatment in these chronically ill schizophrenic patients".

RANDOMISATION

The essential feature of the clinical trial is that the allocation of subjects to one or other of the groups is under the control of the researcher, who tries to ensure that the two groups are as alike as possible, except for the treatment under investigation. However, once a subject has proved eligible to participate in the trial, their allocation to one group or the other should not be influenced by the investigator since this can lead, consciously or unconsciously, to *selection bias*. One way of achieving this is by *randomisation* (or *random allocation*). This has the added virtue that the two groups are also alike in *unknown* as well as known attributes, and if the two groups differ, they will do so purely by chance alone.

One method of randomisation is to use a table of random numbers, such as those in Table A2 in the Appendix at the end of this book. We first make the decision, say, that subjects with odd numbers will be allocated to the treatment group (T) and those with even numbers to the control group (C), or vice versa. We then stick a pin somewhere in the table of random numbers and write down the numbers going down, or across, or up the page. Suppose the numbers are:

    2 4 7 3 4 8 3 5 6 4 1 7 1 9 5 1 . . .

Then the subjects are allocated as follows:

    C C T T C C T T C C T T T T T T . . .

One immediately obvious problem is that we have 10 subjects in the treatment group and only six in the control group. Even with large samples the imbalance can be serious.

MEMO

Randomisation is used in clinical trials to ensure that selection bias is kept to a minimum.

Blocked Randomisation

We can overcome the problem of unequal group sizes by selecting the subjects in successive *blocks*. Suppose we decide on blocks of four. We then write down all possible combinations containing equal numbers of C and T, that is:

1. CCTT   4. TCTC
2. CTCT   5. TCCT
3. CTTC   6. TTCC

Assuming the same set of random numbers as before, 2473483564171951, the first number being 2 means that the first four subjects are allocated to combination 2, i.e. CTCT; the next four to combination 4, TCTC; the third random number, 7, is ignored since we don't have a combination numbered 7; the next four subjects to combination 3, CTTC; and so on, until all subjects have been allocated to either one of the two groups. In this way we will have equal numbers in each group.

If it was important to produce groups containing, say, equal numbers of males and females, we can achieve this by having separate blocks for each sex. This is known as *stratified* block randomisation. Stratification can take place for any possible confounding variables, age being another obvious candidate.

Single-Blind and Double-Blind Designs

Randomisation protects against selection bias. Once subjects have been randomly allocated to either the treatment or the control group, it is also important to insure against *treatment bias*, i.e. the way the trial is actually carried out. This is best achieved if neither the subject nor the researcher knows which subject is receiving which treatment. This ideal trial situation is referred to as a *double-blind* design, and can only be achieved if there is no discernible difference between the treatments given. So for example, in trials to evaluate the efficacy of a new drug, we can't give the control group nothing since this would give the game away. In such cases the control group will be given an identically looking placebo. However, in some circumstances it is impossible for the investigator not to know which treatment is being given, but this knowledge may still be kept from the subject. This sort of design is known as *single-blind*.

An Example from Practice

Researchers[12] conducted a randomised controlled trial, the aim of which was to estimate the changes in size of cardiovascular risk factors in men and women that could be achieved in one year by nurse-led screening and lifestyle interventions. Fifteen towns were selected which met specific demographic criteria and within each town all general practices with four to seven partners were surveyed. From these a pair of willing practices in each town with similar sociodemographic characteristics was randomly allocated either to the screening and intervention group (the treatment group) or to the comparison control group. The treatment group members were screened at the beginning of the trial for height, weight, body mass index, carbon monoxide concentration in breath, blood pressure, and blood cholesterol and glucose. They were given advice on desirable changes in lifestyle and visited regularly during the one-year trial period. The comparison group was screened only at the end of the year.

"Are you sure this is what's meant by a double-blind trial?"

The trial reached 73% (more than 8000) of eligible families. The results found slightly lower weight, blood pressure and blood cholesterol at one year in the intervention group, corresponding to a reduction of "only" 12% in the risk of coronary heart disease events. The authors indicated that this was disappointing given the intensive intervention by trained professionals, and did not augur well for the UK government's voluntary health promotion package for primary care which had no extra financial resources attached to it.

MEMO

Blinding is used in clinical trials to avoid assessment and treatment bias.

SUMMARY

In the chapters preceding this we looked at methods for describing sample data. This chapter has examined a number of ways in which sample data can be gathered. We work with sample data because in most practical situations it is extremely difficult if not impossible to obtain population data (many populations are very large, some are infinite, some difficult to identify).

In an ideal world we would gather sample data by taking simple random samples from the population of interest using tables of random numbers. In the real world with large populations this method would prove too tedious and time-consuming. An alternative is systematic random sampling. If there are particular strata in the population (sex, age, etc.) which we wish to ensure are properly represented in the sample we might use stratified random sampling. These methods require a sampling frame (a list of every member of the population) which for many populations will not be available, in which case a cluster sampling approach may be necessary.

In many situations in health care, random sampling is not a possibility and investigators have to use presenting samples which cannot be viewed as representative of any more general population other than the (restricted) one from which the sample is drawn.

Samples are not, of course, taken in isolation without reference to some investigative study. The designs of such studies take several forms, the type of which will depend on the object of the study and the nature of the framework in which it is set.

Study designs may be either observational (and classified as prospective, retrospective or cross-sectional), or experimental. Prospective or cohort studies follow a group or cohort of individuals forward in time to some end point (death or morbidity) and then determine which members of the cohort were or were not exposed to some risk factor. Retrospective or case–control studies start from some end point and go backwards in time to detect which risk factor individuals (the cases) might have been exposed to in the past. Comparisons are made with members of a control group who have not experienced the end point in question.

Any association between the risk factor and the disease can be assessed using either the relative risk (in the cohort study) or the odds ratio (in the case–control study).

Cross-sectional studies examine samples of individuals (or items) at some moment (or short period) in time, to determine the characteristics and attributes of those in the sample. Cross-sectional studies often use questionnaires to elicit the required information.

A clinical trial is an experimental study design. A new treatment or intervention is given to one group (the treatment group), an existing treatment or placebo to the control group. Randomisation in the allocation of individuals to either group protects against selection bias, blinding protects against treatment or assessment bias.

EXERCISES

7.1 What condition must apply before a sampling procedure can be described as random? Explain why it is often difficult to obtain random samples in health care situations (giving some examples). What are the consequences in terms of identifying the population in question?

7.2 Explain what a sampling frame is. Give some practical examples where it would be difficult to construct a sampling frame. What sampling procedure might you use in such circumstances?

7.3 Why are cohort and case–control studies both often referred to as longitudinal studies? Explain the essential difference between the two designs. What are the principal advantages and limitations of each approach?

7.4 In a cohort study into the control of blood pressure and risk of acute myocardial infarction (AMI) among Swedish subjects[12], the results were analysed for sex and various levels of diastolic and systolic blood pressures. The results for men and women in the risk group of ≥ 180 mmHg systolic blood pressure category were as follows: total number of men = 96; number of AMIs = 9. Total number of women = 206; number of AMIs = 15. In the not-at-risk group, ≤ 84 mmHg systolic blood pressure: total number of men = 227; number of AMIs = 31. Total number of women = 386; number of AMIs = 18. Calculate the relative risk for men and women in the at-risk and not-at-risk groups and compare and interpret your results.

7.5 In an unmatched case–control study of the effects of passive smoking from their husbands on coronary heart disease (CHD)[7], investigators selected a sample of 59 women as cases, 38 of whom had smoking husbands, and 126 women as controls, 58 who had smoking husbands. Calculate the odds ratio and comment on your result.

7.6 In an investigation into the effects of infant exposure to atmospheric sulphur dioxide on the prevalence of bronchial hyper-responsiveness (BHR) among school children aged 12 to 13 years old, researchers used a matched case–control design. The cases were children who had spent their infancy living in a town with an aluminium smelter producing 270–300 kg/h of sulphur dioxide during the study period. The controls were children matched for sex, age, parental smoking habits and socio-economic status, who lived in rural districts free from pollution. The characteristics of the first 12 case–control pairs are shown in Table 7.6. Calculate the odds ratio for these data and interpret your result.

Table 7.6: Case–control pairs from an investigation into bronchial hyper-responsiveness of school children

PAIR	CASES	CONTROLS
1	Y	Y
2	N	N
3	Y	Y
4	Y	N
5	Y	N
6	Y	N
7	N	N
8	N	N
9	Y	N
10	N	Y
11	N	N
12	Y	Y

7.7 Why do you think that non-response to questionnaires in a cross-sectional study can seriously bias any results? Suggest some examples to illustrate your answer.

7.8 What devices are used in clinical trials to overcome the potential problems of (i) selection bias, and (ii) treatment and assessment bias?

REFERENCES

1. Altman, D. G. (1994) *Practical Statistics for Medical Research*. London: Chapman & Hall, 75.
2. Daley, L. E. *et al.* (1991) *Interpretation and Uses of Medical Statistics*. London: Blackwell Scientific, 179.
3. Campbell, M. J. and Machin, D. (1994) *Medical Statistics: A Commonsense Approach*. Chichester: John Wiley, 8.
4. Doll, R. and Peto, R. (1976) Mortality in relation to smoking. *BMJ*, **2**, 1525–36.
5. Groenbaek, M. *et al.* (1994) Influence of sex, age, body mass index, and smoking on alcohol intake and mortality. *BMJ*, **308**, 302–6.
6. Glyn, R. J. *et al.* (1995) Evidence for a positive linear relationship between blood pressure and mortality in elderly people. *The Lancet*, **345**, 825–29.
7. He, Y. *et al.* (1994) Passive smoking at work as a risk factor for coronary heart disease in Chinese women who have never smoked. *BMJ*, **308**, 380–4.
8. Gilbert, R. E. *et al.* (1995) Bottle feeding and the sudden infant death syndrome. *BMJ*, **310**, 89.
9. Bennett, A. E. and Ritchie, K. (1975) *Questionnaires in Medicine: A Guide to their Design and Use*. London: Nuffield Provincial Hospitals Trust.
10. Wright, J. D. (1995) Knowledge and experience of young people regarding drug misuse, 1969–94. *BMJ*, **310**, 20–4.
11. Silver, H. and Geraisy, N. (1995) Effects of biperiden and amantadine on memory in medicated chronic schizophrenic patients: a double-blind cross-over study. *British Journal of Psychiatry*, **166**, 241–3.
12. Lindblatt, U. *et al.* (1994) Control of blood pressure and risk of first acute myocardial infarction: Skaraborg hypertension project. *BMJ*, **308**, 681–6.

APPENDIX

Table A1: Areas under the "standard normal distribution" between $z = 0$ and chosen value of z. The area in the trial is equal to 0.5 minus the value in the table

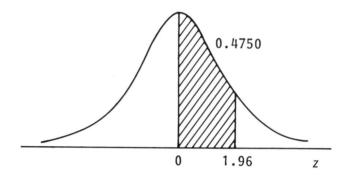

z	.00	.01	.02	.03	.04	.05	.06	.07	.08	.09
0.0	0.0000	0.0040	0.0080	0.0120	0.0160	0.0199	0.0239	0.0279	0.0319	0.0359
0.1	0.0398	0.0438	0.0478	0.0517	0.0557	0.0596	0.0636	0.0675	0.0714	0.0753
0.2	0.0793	0.0832	0.0871	0.0910	0.0948	0.0987	0.1026	0.1064	0.1103	0.1141
0.3	0.1179	0.1217	0.1255	0.1293	0.1331	0.1368	0.1406	0.1443	0.1480	0.1517
0.4	0.1554	0.1591	0.1628	0.1664	0.1700	0.1736	0.1772	0.1808	0.1844	0.1879
0.5	0.1915	0.1950	0.1985	0.2019	0.2054	0.2088	0.2123	0.2157	0.2190	0.2224
0.6	0.2257	0.2291	0.2324	0.2357	0.2389	0.2422	0.2454	0.2486	0.2517	0.2549
0.7	0.2580	0.2611	0.2642	0.2673	0.2704	0.2734	0.2764	0.2794	0.2823	0.2852
0.8	0.2881	0.2910	0.2939	0.2967	0.2995	0.3023	0.3051	0.3078	0.3106	0.3133
0.9	0.3159	0.3186	0.3212	0.3238	0.3264	0.3289	0.3315	0.3340	0.3365	0.3389
1.0	0.3413	0.3438	0.3461	0.3485	0.3508	0.3531	0.3554	0.3577	0.3599	0.3621
1.1	0.3643	0.3665	0.3686	0.3708	0.3729	0.3749	0.3770	0.3790	0.3810	0.3830
1.2	0.3849	0.3869	0.3888	0.3907	0.3925	0.3944	0.3962	0.3980	0.3997	0.4015
1.3	0.4032	0.4049	0.4066	0.4082	0.4099	0.4115	0.4131	0.4147	0.4162	0.4177
1.4	0.4192	0.4207	0.4222	0.4236	0.4251	0.4265	0.4279	0.4292	0.4306	0.4319
1.5	0.4332	0.4345	0.4357	0.4370	0.4382	0.4394	0.4406	0.4418	0.4429	0.4441
1.6	0.4452	0.4463	0.4474	0.4484	0.4495	0.4505	0.4515	0.4525	0.4535	0.4545
1.7	0.4554	0.4564	0.4573	0.4582	0.4591	0.4599	0.4608	0.4616	0.4625	0.4633
1.8	0.4641	0.4649	0.4656	0.4664	0.4671	0.4678	0.4686	0.4693	0.4699	0.4706
1.9	0.4713	0.4719	0.4726	0.4732	0.4738	0.4744	**0.4750**	0.4756	0.4761	0.4767
2.0	0.4772	0.4778	0.4783	0.4788	0.4793	0.4798	0.4803	0.4808	0.4812	0.4817
2.1	0.4821	0.4826	0.4830	0.4834	0.4838	0.4842	0.4846	0.4850	0.4854	0.4857
2.2	0.4861	0.4864	0.4868	0.4871	0.4875	0.4878	0.4881	0.4884	0.4887	0.4890
2.3	0.4893	0.4896	0.4898	0.4901	0.4904	0.4906	0.4909	0.4911	0.4913	0.4916
2.4	0.4918	0.4920	0.4922	0.4925	0.4927	0.4929	0.4931	0.4932	0.4934	0.4936
2.5	0.4938	0.4940	0.4941	0.4943	0.4945	0.4946	0.4948	0.4949	0.4951	0.4952
2.6	0.4953	0.4955	0.4956	0.4957	0.4959	0.4960	0.4961	0.4962	0.4963	0.4964
2.7	0.4965	0.4966	0.4967	0.4968	0.4969	0.4970	0.4971	0.4972	0.4973	0.4974
2.8	0.4974	0.4975	0.4976	0.4977	0.4977	0.4978	0.4979	0.4979	0.4980	0.4981
2.9	0.4981	0.4982	0.4982	0.4983	0.4984	0.4984	0.4985	0.4985	0.4986	0.4986
3.0	0.4987	0.4987	0.4987	0.4988	0.4988	0.4989	0.4989	0.4989	0.4990	0.4990

Table A2:   Random numbers

23157	54859	01837	25993	76249	70886	95230	36744
05545	55043	10537	43508	90611	83744	10962	21343
14871	60350	32404	36223	50051	00322	11543	80834
38976	74951	94051	75853	78805	90194	32428	71695
97312	61718	99755	30870	94251	25841	54882	10513
11742	69381	44339	30872	32797	33118	22647	06850
43361	28859	11016	45623	93009	00499	43640	74036
93806	20478	38268	04491	55751	18932	58475	52571
49540	13181	08429	84187	69538	29661	77738	09527
36768	72633	37948	21569	41959	68670	45274	83880
07092	52392	24627	12067	06558	45344	67338	45320
43310	01081	44863	80307	52555	16148	89742	94647
61570	06360	06173	63775	63148	95123	35017	46993
31352	83799	10779	18941	31579	76448	62584	86919
57048	86526	27795	93692	90529	56546	35065	32254
09243	44200	68721	07137	30729	75756	09298	27650
97957	35018	40894	88329	52230	82521	22532	61587
93732	59570	43781	98885	56671	66826	95996	44569
72621	11225	00922	68264	35666	59434	71687	58167
61020	74418	45371	20794	95917	37866	99536	19378
97839	85474	33055	91718	45473	54144	22034	23000
89160	97192	22232	90637	35055	45489	88438	16361
25966	88220	62871	79265	02823	52862	84919	54883
81443	31719	05049	54806	74690	07567	65017	16543
11322	54931	42362	34386	08624	97687	46245	23245

INDEX

Wiley Titles of Related Interest...

RESEARCH IN HEALTH CARE
Design, Conduct and Interpretation of Health Services Research
I. K. CROMBIE with H.T.O. DAVIES

An invaluable and practical text which provides health care professionals with a comprehensive and readable guide to the design, conduct and interpretation of health services research. It strips the research process of its technical jargon, illuminating the methods of research by an extensive use of real examples.

0471 96259 7   312pp   1996   Paperback

MEDICAL STATISTICS
A Commonsense Approach
Second Edition

M. J. CAMPBELL and D. MACHIN

Written by experts with wide teaching and consulting experience, the second edition of this successful book presents a concise explanation of the principles of medical statistics. Examples are genuine and taken from the authors' own research.

0471 93764 9   200pp   1993   Paperback

ADEQUACY OF SAMPLE SIZE IN HEALTH STUDIES
S. LEMESHOW, D.W. HOSMER Jr., J. KLAR and S.K. LWANGA

Provides epidemiologists and health workers with a good basic knowledge of sampling principles and methods and an acquaintance with their potential applications in the medical field.

0471 92517 9   252pp   1989